Luciana Dantas dos Santos
Roberto Lima Barcellos

Current Sedimentation of the Rio Formoso Estuary - PE (Brazil)

Luciana Dantas dos Santos
Roberto Lima Barcellos

Current Sedimentation of the Rio Formoso Estuary - PE (Brazil)

Sedimentology and seasonal geochemistry in a Brazilian tropical estuarine system

ScienciaScripts

Imprint
Any brand names and product names mentioned in this book are subject to trademark, brand or patent protection and are trademarks or registered trademarks of their respective holders. The use of brand names, product names, common names, trade names, product descriptions etc. even without a particular marking in this work is in no way to be construed to mean that such names may be regarded as unrestricted in respect of trademark and brand protection legislation and could thus be used by anyone.

Cover image: www.ingimage.com

This book is a translation from the original published under ISBN 978-3-330-76330-2.

Publisher:
Sciencia Scripts
is a trademark of
Dodo Books Indian Ocean Ltd. and OmniScriptum S.R.L publishing group

120 High Road, East Finchley, London, N2 9ED, United Kingdom
Str. Armeneasca 28/1, office 1, Chisinau MD-2012, Republic of Moldova, Europe
Managing Directors: Ieva Konstantinova, Victoria Ursu
info@omniscriptum.com

Printed at: see last page
ISBN: 978-620-8-40265-5

Index

Authors

MSc Luciana Dantas dos Santos: MSc in Geosciences from the Postgraduate Programme in Geology at UFPE (2016). Graduated in Geography from IFRN (2012). She works on studies of marine sedimentology with an emphasis on environmental geochemistry, sediment dynamics and sedimentary organic matter in coastal and estuarine areas. Contact: santos.luciana27@yahoo.com.br

PhD. Roberto Lima Barcellos: BA in Geography from the University of São Paulo (1996). Master's (2000) and PhD (2005) in Chemical and Geological Oceanography from the Postgraduate Programme in Oceanography at the Oceanographic Institute of USP. Post-doctorate from PPGO-IOUSP (2009). He is currently leader of the CNPq research group: Study Group in Sedimentology and Marine Geochemistry (GSGMar- DOCEAN-UFPE) and Adjunct Professor III at the Department of Oceanography at UFPE. He works on research into Geological Oceanography, with an emphasis on marine sedimentology and geochemistry, spatio-temporal morphosedimentary dynamics, stable isotopes and pollution. Contact: roberto.barcellos@ufpe.br

Thank you

The authors would first like to thank the support of the Marine and Tropical Ichthyology Group (IMAT) of the Department of Oceanography of the Federal University of Pernambuco (DOCEAN-UFPE), which coordinated this research through Prof. Dr Maria Elisabeth Araújo, whose role was essential for the preparation and realisation of this book from the field data collection activities.

To the Geological Oceanography Laboratory (LABOGEO-DOCEAN) and the Marine Sedimentology and Geochemistry Study Group (GSGMar-DOCEAN) for their support in field and laboratory activities, in helping to carry out the sedimentological and geochemical analyses, as well as in the statistical treatment of the data generated by the work.

To the support of the Chemical Oceanography Laboratory (LOQUIM) coordinated by Prof. Dr Manuel de Jesus Flores Montes (DOCEAN) in the analyses of sedimentary phosphorus. And the Laboratory of Organic Compounds in Coastal and Marine Ecosystems (ORGANOMAR) for the collaboration of Prof Gilvan Takeshi Yogui (DOCEAN).

To CNPq for their financial support for the research project "Health of reef ecosystems subject to different intensities of use and conservation in Suape and Tamandaré". And to the boatman *Seu Neco* for his in-depth knowledge of the estuarine channels of the Rio Formoso and who helped us a lot with the sediment collection strategy in the field.

To the Department of Oceanography at UFPE for allowing the author and LABOGEO technician MSc. Luciana Dantas dos Santos to study for her master's degree in the Postgraduate Programme in Geosciences (PPGEOC-UFPE), for which we are equally grateful for their support in the development of the research. And finally, the authors would like to thank their respective families for their love, encouragement and support in carrying out the research translated into this work.

Foreword

Coastal zones represent the dynamic intersection between the different physical environments that make up planet Earth. This unique congruence between the lithosphere, hydrosphere, atmosphere, biosphere and eventually the mesosphere cannot, however, be considered just as a simple line, in this case the coastline, as it is represented cartographically.

In fact, the coastal zone is a portion of the planet full of complex processes associated with its geological heritage, sediment flow and relative sea level fluctuations. It is thus spatially and physically represented by the origin and evolution of the coastal sedimentation environments that make it up, which are extremely dynamic and ephemeral on the scale of Geological Time.

Among the various sedimentary environments that make up the coastal zone, estuaries stand out. The term estuary comes from the Latin *aestus* meaning heat, boiling or tide, and the adjective *aestuarium* specifically means tide. Estuarine systems represent the transition between fluvial and marine environments and their origin and evolution is mainly related to the rise in mean sea level and the drowning of river valleys. They are typical of transgressive coastlines, such as the south coast of Pernambuco in the north-east of Brazil.

Estuaries act as one of the main conduits for sediment from continental sources to the oceans, being sites of rapid sedimentation. They act as a filter and trap for materials, especially fine sediments, nutrients and organic matter. These are retained in their domains, resulting in high primary productivity. They also act as a shelter and nursery for marine and transitional species. They are most often found in coastal regions subject to meso- and macrotidal regimes, although they are present in coastal zones all over the world, regardless of climate. River flow, the action of gravity waves in its more oceanic portions and, in particular, tidal dynamics are the hydrodynamic forces that give each estuarine system heterogeneous and particular sedimentary characteristics, depending on the weight of each conditioning factor and the adjustment to the bottom topography, which comes from geological heritage. The relationship between estuarine environments and human beings is an ancient one, in which prehistoric man, associated with lake environments, reached the oceans via estuaries. And the estuarine vocation as a shelter for harbours, bases for trade and an easy and abundant source of food, led humans to occupy and settle in these areas. Thus, the concentration of people and wealth around such environments has led to the search for more available land along the coast, in coastal systems that interact in a delicate balance between opposing forces such as the rate of siltation and wave erosion. And this extensive use of estuaries has also relegated them to a dumping ground for waste from the surrounding developing communities. Today, major cities and several of the world's megalopolises are located on the shores of estuaries, a direct reflection of the fact that 70% of the population is within 200 kilometres of the coastline.

However, there is an environmental and economic paradox associated with estuarine environments. The same oceanographic conditions that were decisive for the construction of these cities in sheltered waters often lead to a very slow capacity to disperse the waste and pollutants dumped in these environments, which are often at a high stage of degradation.

Human presence in estuarine environments is historical and global in nature, generating imbalances and strong environmental pressure on these fragile and important sedimentation sites. In particular, they generate conflicts over land use, disorderly urban occupation, alteration of the banks of estuarine channels, destruction of mangroves/marshes/swamps and the removal of living and non-living resources, with the consequent imbalance of fauna and flora. This set of factors often results in an imbalance in the sediment balance of estuarine systems, causing often irreversible processes of erosion, clogging and silting, with incalculable damage to public and private managers and the population in general.

In the current context of climate change, with IPCC data indicating a current (2017) rise in sea level of 4.2mm/year, we are led to believe that these environments will suffer major impacts in the 20th century. And in coastal zones dominated by estuarine environments that have already been suffering changes in the sediment balance in recent decades, as well as interference from coastal works, such as on the coast of the state of Pernambuco (NE-Brazil), these forecasts become more critical.

The southern coast of the state, in turn, is part of the Zona Mata Sul region and has the Formoso river estuary as a prominent coastal river. Its heritage and tectonic conditioning have led to its current physiographic compartmentalisation of great scenic beauty. Since the 2000s, this region has seen a great deal of development in tourism, agriculture and extractive activities, which will certainly increase the pressure on local coastal systems such as the estuary that is the subject of this book.

Urban sprawl, associated with tourism and property speculation, has been growing steadily, regardless of the fact that the estuarine region is located on the northern edge of the APA Costa dos Corais (Federal Conservation Area - ICMBio) and is entirely within the state APA of Guadalupe (PE). However, being an integral part of environmental protection areas has not prevented the occurrence of problems related to various environmental impacts. The area is currently under strong pressure from the tourism industry, real estate, carcinoculture and polycultures, including sugar cane cultivation.

In view of the above, the aim of this book is to evaluate seasonal sediment distribution and the behaviour of sedimentary organic matter as environmental indicators in the Formoso River estuarine system. To correlate all the parameters studied, such as granulometry, total organic matter content, calcium carbonate and elemental carbon and sedimentary phosphorus contents. And from the evaluation of these geochemical parameters, we can obtain information on the seasonal sedimentary behaviour in the estuary, associating them with possible local environmental impacts, derived mainly from carcinoculture and sugar cane activities. The results of this research are summarised in 7 chapters: Introduction; Study site; Seasonal sedimentology of the Rio Formoso estuary; Environmental geochemistry; Statistical analysis of the parameters: sedimentology and geochemistry; Integrated discussion of the data; and Final summary.

It should also be noted that this research essentially fulfils the objectives and is an integral part of the project: "Health of reef ecosystems subject to different intensities of use and conservation in Suape and Tamandaré" (CNPq), which is a partnership between the Sedimentology and Marine Geochemistry Study Group (GSGMar-DOCEAN) and the Marine and Tropical Ichthyology Group (IMAT-DOCEAN).

Finally, it is hoped that this work will contribute to the knowledge of seasonal sedimentary processes and environmental geochemistry in similar estuaries throughout the world's tropical coastal zones.

Roberto Lima Barcellos

Recife, 04 May 2017.

CHAPTER 1

Introduction

The Pernambuco coastline has low altitudes, reaching altitudes below the high tide level at several points (MANSO et al., 2006). This fact justifies the creation of a fluvial-marine environment, since it allows Atlantic waters to enter the coastal relief, which favours the emergence of 13 estuaries on the Pernambuco coast (SILVA et al., 2009).

The term estuary comes from the Latin *aestus* meaning heat, boiling or tide. Specifically, the adjective *aestuarium* means tide. The most widely used definitions of estuaries in scientific literature are given by Pritchard (1967): "An estuary is a semi-enclosed coastal body of water which has a free connection with the sea and in which seawater is diluted measurably with fresh water from land drainage." And by Fairbridge (1980): "An estuary is a seawater inlet within a drowned river valley up to the upper tidal limit, usually being divided into three sectors (a marine or lower estuary, in connection with the open sea; a middle estuary, subject to a strong mixture of salt and fresh water; and an upper or river estuary, characterised by fresh water but subject to daily tidal action)."

As estuaries are considered coastal and transitional environments, they receive energy and material from both terrestrial and marine systems (AGUIAR, 2005). In this way, they play a fundamental role in the sediment cycle. It promotes exchanges between the oceans and the continents, often acting as favoured areas of deposition. The continuous mixing of fresh waters with those of higher salinity presents physiological problems for estuarine plants and animals that are not adapted to this environment. The suspended material brought in by the rivers and accumulated on banks produces areas rich in food for many organisms, but on the other hand causes low oxygenation or even anoxia (TUNDISI & TUNDISI, 2008).

Considering sediment as the compartment that reflects all the processes taking place in an aquatic ecosystem, its composition should also give an indication of its state of trafficking. The concentration of some sediment components reflects the level of production in the system, as is the case with organic matter (ESTEVES, 1998). According to Naumann (1930 *apud* Esteves, 1998), in oligotrophic lakes the sediment is characterised by low organic matter content and low nutrient concentration. On the other hand, in mesotrophic lakes and especially in eutrophic lakes, the organic matter content and nutrient concentration increase considerably.

Studies on sedimentary organic matter thus make it possible to assess the environmental conditions of sedimentary areas, providing analyses of their natural conditions and inferences about anthropogenic action in them. The association of organic matter concentrations with sediment distribution, with the origin of the organic material and the oxidising conditions of the bottom, makes it possible to understand the current sedimentation process and the different factors that interfere in the process, both in time and space (BARCELLOS, 2005).

According to Pettijohn (1975), granulometric parameters are good tools for interpreting the hydrodynamics of the bottoms of marine areas. The same author also states that the accumulation of organic matter in sediments is strongly dependent on the amount of clay deposited, due to the adsorption process. Trask (1939 *apud* Tyson, 1995) also adds that the organic content of continental margin sediments generally increases when the grains that make them up become finer. This is because clays commonly have about twice as much organic matter as silt, and about four times as much organic matter as very fine sands. The organic content can therefore be directly correlated with the median, mean diameter and, above all, the percentage of clay in the sediment.

1.1 Elemental composition of organic matter: C and P

1.1.1 Organic Carbon (C)

According to Esteves (1998), the types of organic carbon found in an aquatic ecosystem can be grouped into two categories: detrital organic carbon and biota organic carbon (COP-biota), which together make up total organic carbon (TOC). Detrital organic carbon, in turn, is made up of two fractions: dissolved organic carbon (DOC) and detrital particulate organic carbon (COP-detrital). According to Martinelli (2009), the main forms of carbon in aquatic systems are: CID (dissolved inorganic carbon), COD (dissolved organic carbon), COP (particulate organic carbon) and COB (organic carbon in biota), whose main reservoir is the oceans.

The smallest reservoir of these forms of carbon in the ocean is biota, of which the main carbon input is photosynthesis (primary production). Another form of carbon input into the oceans is the transport of organic carbon by rivers. Rainwater entering the soil is further enriched with CO2 and reacts with carbonates. Rivers are one of the main inputs of carbon from the continent to the coastal ecosystem. Its composition depends on the type of vegetation, the soil and the lithology of the drainage basin (Souza et al., 2012). In systems located near highly populated and industrialised regions, domestic sewage, industrial effluents and drainage from urban areas contribute significantly to carbon input, mainly in organic form. These anthropogenic sources can eventually become more important than natural inputs.

The carbon content of surface sediments depends on a number of factors, such as sedimentary characteristics, water column productivity, the rate of microbial degradation, as well as local oceanographic conditions. According to Huc (1980, *apud* Rashid, 1985), a low organic carbon content (< 0.5%) is characteristic of the vast majority of ocean basins, particularly those in open seas. Sediments close to the coastline, inland seas and continental shelves are generally enriched in organic carbon. Contents of 2 to 4 per cent are not uncommon in these areas.

1.1.2 Phosphorus (P)

The phosphorus cycle in continental aquatic systems has an important component in the sediments. Part of the phosphorus undergoes a process of complexation during periods of intense sediment oxygenation and thus becomes periodically unavailable. As phosphorus does not have a gaseous component, its availability depends on phosphate rocks and the internal cycle of estuaries, of which decomposition and the excretion of organisms are important parts (TUNDISI & TUNDISI, 2008).

The phosphate present in continental aquatic ecosystems comes from natural and artificial sources. Among the natural sources, drainage basin rocks are the basic source of phosphate. It comes from the weathering of primary phosphate minerals in continental rocks. The most important of these is apatite (ESTEVES, 1998). Among the artificial sources, anthropogenic influence stands out. Among these, the phosphate source can be distinguished by leaching from farmland soils and sewage emissions (in the form of detergents, human and industrial waste). The orthophosphate ions that are released into the environment are solubilised through the leaching process carried out by rainfall and subsequently reach river courses. Rivers are then the predominant means of transferring continental phosphorus to the oceans (Riley & Chester, 1978). In this way, marine environments adjacent to areas subject to intense agricultural activity or sewage emissions can have significantly higher concentrations of this element in sedimentary organic matter.

According to Filippelli (1997) the phosphorus concentrations found in sediments from continental margins with high sedimentation rates vary from 8 to 108 pmol/g, for ocean basins from 7 to 307 pmol/g and for phosphogenic environments from 580 to 3700 pmol/g. For non-phosphogenic continental margins, the values found were: 92 to 108 pmol/g (W Africa), 35 to 80 in/g (Peru), 23 to 33 pmol/g (California) and 23 to 24 pmol/g (North Carolina). In coastal sediments, such as the Mississippi River Delta and Long Island Bay in the United States (Ruttenberg & Bemer, 1993 *apud* Filippelli, 1997) and the St Lawrence River Estuary in Canada (Sundby *et al.* 1992 *apud* Filippelli, 1997), the average values varied around: lopmol/g, 8|umol/g and 26-32pmol/g, respectively.

The following levels have been found for other tropical and sub-tropical estuarine environments around the world: Bertioga Channel (SP) (7.0 to 38.1 pmol/g) (MAHIQUES et al., 1997), Santos-São Vicente Estuary (0.3 to 80.6 pmol/g) (SIQUEIRA, 2003), Cananéia-Iguape estuarine lagoon system (1.44 to 39.75 pmol/g) (BARCELLOS, et al., 2009), Guanabara Bay (RJ) (40 pmol/g) (CARREIRA & WAGENER, 1998), 2009), Guanabara Bay (RJ) (40 pmol/g) (CARREIRA & WAGENER, 1998), Tolo estuary (Hong Kong) (14.5 pmol/g) (THOMPSON & YEUNG, 1994 *apud* CARREIRA & WAGENER, 1998), Culiacan estuary (Mexico) (13 to 97 pmol/g) (RUIZ-FERNÁNDEZ et al., 2002), estuaries from Texas to Florida (USA) (1.8 to 45.9 pmol/g) (HUANXIN et al., 1994), Yangtze River estuary (China) (18.0 to 31.4 pmol/g) (XU et al., 2001). Ganges Delta- Brahmaputra- Meghna (India-Bangladesh) (16.1 to 58.1 pmol/g) (DATTA et al., 1999).

On the coast of Pernambuco, Gaspar (2008) found values of up to 47.03 pMol/g of total phosphorus for the Botafogo River, 45.52 pMol/g for the Santa Cruz Channel and 38.30 pMol/g for the Carrapicho River.

1.2 C/P ratio as an indicator of material source

The C/P Ratio can be used to determine the predominance of continental or marine contributions to the organic matter present in the sediments (table 1).

Table 1: Elementary C/P ratio and its interpretation

Reason	Origin of Organic Matter	Reference
C/P	7-80 Microbial activity	Ruttenberg & Goni, 1997
	80-300 Marine or mixed origin	
	300 - 1300 Soft plant tissues	
	> 1300 Hard Plant Tissues	

The two major sources of organic matter for marine sediments are terrestrial and marine plants. These organisms have different C/P ratios. Marine phytoplankton have an average molar C/P ratio of 106 (Redfield et al., 1963 *apud* Ruttenberg & Goni, 1997), in contrast to higher plants which, according to various authors, are relatively depleted in phosphorus, with C/P ratios ranging from 300 to 1300 for soft tissues and ratios greater than 1300 for hard tissues (Ruttenberg & Goni, 1997). C/P ratios indicating marine or mixed origin range from 80 to 300 (Ruttenberg & Goni, 1997). Microbial communities can also be important components of sedimentary organic matter. The C/P ratios of bacteria range from 7 to 80 (Gãchter & Meyer, 1993 *apud* Ruttenberg & Goni, 1997).

Ruttenberg & Goni (1997) suggest that for coastal areas, where the input to the sediments is characterised by a mixture of marine and terrigenous phytodetritus, higher values of the C/P ratio (above the Redfield Ratio, i.e. >106) may indicate that the sediments are enriched in organic matter of terrigenous origin, while values close to the Redfield Ratio would indicate organic matter of predominantly marine origin.

1.3 Objectives of the book, materials and research methods

The general aim of this book is to analyse the sedimentary process in the estuarine region of Rio Formoso (PE), with the aim of understanding the seasonal dynamics of sedimentological and geochemical processes in the estuary under study. Its specific objectives are: 1) Analyse the compositional characteristics of organic matter and sediments in the Formoso River estuary (PE); 2) Determine the spatial distribution of the two elements studied (TOC and P), analysing the distribution and origin of organic matter on a seasonal basis; 3) To relate the concentrations of TOC and P to the distribution of sediment and the sedimentation process of natural and anthropogenic origin; 4) To use sedimentological parameters as environmental indicators; 5) To assess how the distribution of sediment and organic matter is affected by the joint action of certain agents and factors, such as: physiography of the region, hydrodynamics (tides and currents) and hydrographic system; 6) Identify areas of possible accumulation of pollutants based on the sedimentological data generated, as well as areas subject to current impacts.

In order to achieve the objectives, some of the methodological procedures necessary to complete the research were carried out, divided into four distinct stages: the initial stage, the field stage, the laboratory stage and the office stage.

1.3.1 Initial stage

In the initial stage, a bibliographical survey was carried out of studies carried out in the coastal zone of Pernambuco, focusing on those whose area of study is the Rio Formoso Estuary Region. With a focus on those that deal with the geochemistry of estuarine sediments and the origin of organic matter in estuaries. This survey was carried out throughout the research period. It consisted of understanding and interpreting scientific articles, master's dissertations and doctoral theses, as well as articles published in national and international journals.

This was followed by a survey of figures, photographs, maps and tables of the study area and the selection of images to be included in this dissertation.

Finalising the initial stage, the field stage was planned and scripted, with the period for carrying out these activities defined.

1.3.2 Field stage

The field stage comprises sedimentological sampling activities.

1.3.2.1 Sedimentological sampling

Twenty-four sampling points were selected for the collection carried out in September (25/09/2014), characterising the end of the rainy season for the region. The same twenty-four points were selected for sampling at the end of the region's dry season in March (23/03/2015) (figure 1).

The surface layer of bottom sediments in the estuary was collected at each sampling point using a Van Veen-type dredger, a sediment tray and a stainless steel spoon on the banks and in the estuary channel.

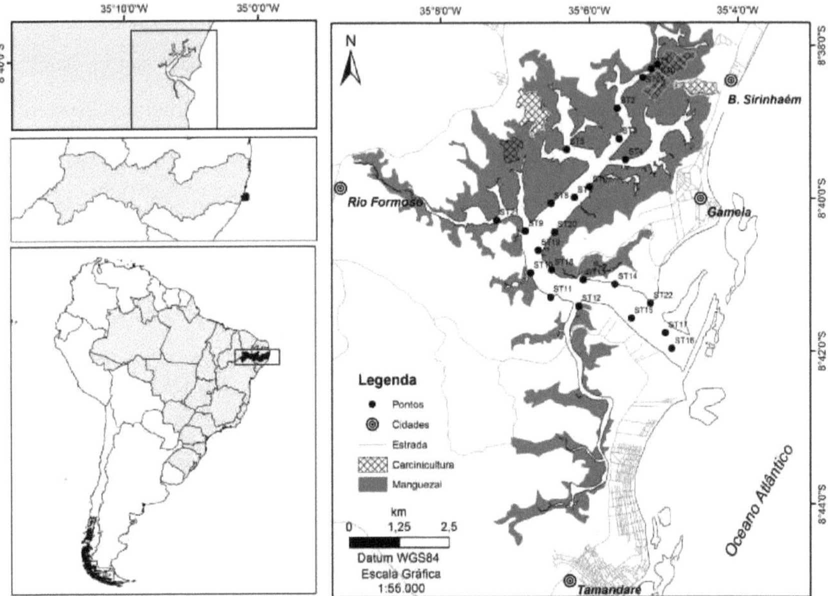

Figure 1: Distribution map of the sampling points. Source: Diego Xavier - LABOGEO/UFPE (2016).

For the analyses of phosphorus (P) and total organic carbon (TOC), around 20g of the sediment samples were stored in ziplock bags with hermetic seals measuring 10x6 cm, preserved in ice until the moment of preparation for the analysis in order to avoid bacterial activity after collection. For sediment granulometric characterisation, total organic matter (TOC) analysis and carbonate (CaCOs) analysis, around 25 Og were packed in 20 x 30 cm plastic bags, tied and stored at room temperature.

1.3.3 Laboratory stage

1.3.3.1 Sample preparation for analyses

The procedure for preparing sediment samples from the Rio Formoso estuary for sediment granulometric characterisation, MOT analysis and CaCO.3 analysis was carried out at the Geological Oceanography Laboratory (LABOGEO) of the Oceanography Department at UFPE. The samples were dried in an oven at 60°C. After drying, they were subdivided into aliquots: 30 g for MOT analysis and later particle size analysis and 10 g for carbonate content analysis.

Sample preparation for phosphorus (P) and total organic carbon (TOC) analysis consisted of thawing, drying and separating aliquots: 0.2g for phosphorus analysis and 0.2g (macerated) for total organic carbon (TOC) analysis. The phosphorus content was analysed at the Chemical Oceanography Laboratory (LOQUIM) of the Oceanography Department at UFPE.

1.3.3.2 MOT and CaCOi content analysis

Determination of carbonate content and total organic matter determined from the difference in dry weight, before and after attack with a 10% HC1 solution and a 10% H2O2 solution, respectively (CARVER, 1971).

1.3.3.3 Particle size analysis

The technique used in the granulometric characterisation of the sediments consists of separating the main textural classes of coarse and fine sediments, following the methods described in Suguio (1973). For coarse sediments (fraction > 0.062mm), an arrangement of sieves according to the Wentworth (1922) aperture was used, at half intervals of phi (o). For fine-grained sediments (fraction < 0.062 mm), the pipetting method based on Stokes' sedimentation law was used, adopted only in cases where the fraction of fines was greater than 10% of the weight of the total sample.

1.3.3.4 Compositional analysis of the sandy fraction and grain moifoscopy

The sediments were analysed under a binocular magnifying glass, then, after observing the entire sample, approximately 200 grains were counted and identified, based on a modification of Hubert's (1971) proposal for both analyses. According to Mahiques et al. (1998) the compositional analysis of the sandy fraction can be used to characterise sub-environments in coastal areas. The first step in analysing the data was to determine an index that could provide a measure of the marine biogenic influence on each sample. This index is known as the marine biogenic index (BM) and is defined by the difference between the relative frequency of marine and terrigenous biogenic constituents. The BM index is a measure of the trend of marine biogenic influence on a given sample, ranging from (-1.0) to (1.0) so that values closer to (-1.0) are related to a greater contribution of continental constituents (MAHIQUES et al., 1998).

For the morphoscopic analysis of the grains, the degree of roundness and (visual) sphericity were inferred by classifying them according to the visual roundness estimation chart by Krumbein; Sloss (1963). The data obtained for the fractions analysed, retained in the 0.707-0.500mm (10) and 0.350-0.250mm (20) intervals, were interpreted separately. These particle size classes, when subjected to the action of currents, undergo different types of transport, in this case traction and saltation, respectively.

1.3.3.5 Total Organic Carbon Analysis

The determination of total organic carbon (TOC) in sediments was based on the method proposed by Gaudette, Muller & Storffers (1974), which is a modification of the Walkley & Black method (1934). This method is based on the oxidation of dissolved and particulate organic material with potassium dichromate to an exact known volume, in a medium of excess concentrated sulphuric acid. The organic material is oxidised by sulphochromic mixing, using an appropriate redox indicator, and the excess oxidant (potassium dichromate) is dosed by a solution of ferrous sulphate O.1N. The titrating solution is standardised with glucose and the organic carbon is measured as glucose carbon (CóH^Oô). Finally, the concentration of total organic carbon [%C_{org}] is determined using:

$$[\%C_{org}] = V.\left(\frac{1-T}{TBr}\right).k.N.\left(\frac{100}{m}\right)$$

Where *V is* the volume of dichromate used (ml); *N is* the normality of the dichromate; T is the volume of ferrous sulphate used in the sample (ml); Tg_r *is* the volume of ferrous sulphate used in the blank (ml); *m* is the weight of the sediment sample (g); *k* = 12/4000 = 0.003 = milliequivalent by weight of carbon.

1.3.3.6 Sedimentary Phosphorus Analysis

The chemical fractions of phosphorus in the sediment were analysed using the modified method proposed by Willians et al. (1976), described by Pardo et al. (2004). This is the method recommended by the European Commission's "Standards, measurements and testing (SMT) programme". The SMT sediment standard BCR 684 was used for calibration and validation of the methodology.

The determination of phosphorus (P) in sediment, organic and inorganic, followed the method proposed by Willians et al. (1976) modified, described by Pardo et al. (2004). This is the method recommended by the European Commission's "Standards, measurements and testing (SMT) programme". In this procedure, only a 0.2g aliquot of the sample is used, which is then digested in a 1 M HC1 solution under constant stirring for 16 hours. The total inorganic phosphorus (TIP) is determined from the extract. After digestion, the residue is calcined for one hour in a muffle furnace at 450°C. It is then digested again in a 1 M HC1 solution under constant stirring for 16 hours. From this new extract, total organic phosphorus (TOP) is determined. At the end of the different extractions, the method measures phosphorus with ammonium molybdate ((NH^óMovCb-i.íEO) in an acidic medium, forming a complex that is reduced by ascorbic acid, resulting in a blue-coloured compound whose maximum absorption is at 885 nm. The concentration is determined against a standard curve of potassium phosphate KH2PO4 by spectrophotometry in the visible region, using the molybdate blue methodology and successive dilutions of SMT's BCR 684 reference standard solution to calibrate the equipment. The detection and quantification limits of the methodology were 0.07 mg/1 and 0.23 mg/1 respectively, with precision of 0.8% and accuracy of 97.0%. The concentration of total phosphorus (TP) was obtained by adding the concentrations of POT and PIT.

1.3.4 Cabinet stage

1.3.4.1 Calculating the C/P ratio

The calculation of the C/P ratio was based on the relationship of Redfield et al. (1963), which divides the percentage of total organic carbon by the percentage of total organic phosphorus (POT).

1.3.4.2 Sysgran

The data resulting from the granulometric analysis was processed according to the statistical parameters of Folk & Ward (1957) and Shepard's (1954) textural facies diagram using the *Sysgran* (3.1) programme (Camargo, 2006), which enabled the average diameter, degree of selection, asymmetry, kurtosis and the content of granules, sand, silt, clay and mud (silt + clay) to be obtained.

1.3.4.3 Statistical treatment of the data obtained

For the joint treatment of data from the rainy/2014 and dry/2015 collection seasons, the Folk & Ward (1957) statistical parameters and Shepard's (1954) triangular diagram were analysed using the *Sysgran* (3.1) programme. Cluster analysis and principal component analysis (PCA) were also carried out using *Primer 6.0* software. The parameters used for the cluster and PCA analyses were: gravel fraction (%), sand fraction (%), mud fraction (%), CaCO content (%), MOT (%), TOC (%), PT (%), POT (%) and PIT (%). To compare spatial variability and determine independent relationships between random variables, *Sperman*'s non-parametric correlation coefficient *(rs)* was used using the *Statistica 13* programme.

1.3.4.3.1 Spearman's correlation (rs)

The *Sperman* coefficient *(rs)* varies between (-1) and (1). The closer the values are to these extremes, the greater the association between the variables. The negative sign of the correlation means that the variables vary in the opposite direction, i.e. that the categories with higher values of one variable are associated with the categories with lower values of the other variable (FERNANDES, 1999).

1.3.4.3.2 Principal component analysis (PCA)

This analysis was used to determine the sedimentological and geochemical parameters that best demonstrated the environmental variability found in the study area.

1.3.4.3.3 Cluster analysis and faciological analysis using a multivariate technique

For the joint treatment of the data and the faciological analysis, they had to be transformed using the expression [$\log_{10}$(x 4-1)] to bring the data series closer to the distribution and normalise them (REGAZZI, 2000). The "Euclidean distance" was the distance index adopted and the grouping analysis method was the "Unweighted Average" (UPGMA).

This clustering technique is designed to produce groups of objects, in this case sedimentary facies, which correspond to recognising a certain degree of similarity between them, sufficient to be able to bring them together in the same group (Romesburg, 1984).

The Simple Euclidean Distance was used because it is an index that makes it possible to measure the distance between two objects, in this case the characteristics of the sedimentary facies, when they are observed as points in the two-dimensional space formed by their attributes. Groups were defined from the two dendrograms generated, arbitrarily using the Euclidean distance cut-off value of 4.0. The characterisation of the samples according to the groups, for the rainy/2014 and dry/2015 periods, was plotted on maps, allowing the spatial characterisation of the distinct facies for the surface sediments of the two periods studied and the creation of sedimentary facies maps for the area under study.

1.3.4.4 Preparation of maps, graphs and tables

The location map of the study area and the distribution maps for the spatial analysis of the parameters used were drawn up using the ArcGISLO programme, while the graphs and tables were drawn up using the *Excel for Windows 16* programme.

CHAPTER 2

Study site

2.1 Location of the study area

The Formoso River estuary covers an area of approximately 27 km^2 and is located on the south coast of Pernambuco between parallels 08°39' and 08°44' South latitude and 35°10' and 35°06' West longitude. It is part of the Mesoregion of Mata Pernambucana, the Microregion of Mata Meridional Pernambucana and the development region of Mata Sul, bordered to the north by the municipality of Sirinhaém, to the south by the municipality of Tamandaré and to the east by the municipality of Rio Formoso (CONDEPE, 1992) (Figure 2). The estuary is around 100 km from Recife and can be accessed via the BR-101 and PE-060 motorways, from which the PE- 061, PE-073 and PE-076 motorways can be taken (SETUR/CPRH, 2011).

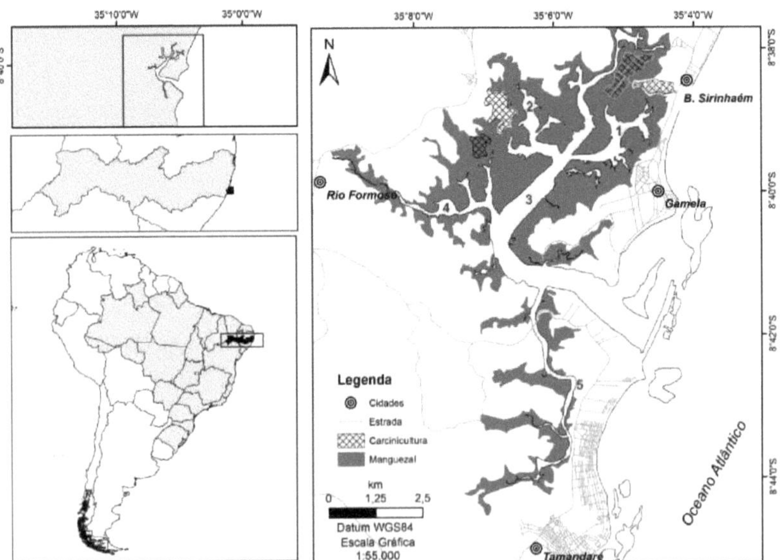

Figure 2: Location map of the study area: Rio Formoso Estuary. A) 1- Lemenho River; 2- Porto das Pedras River; 3- Passos River; 4- Formoso River; 5- Ariquindá River. Source: Diego Xavier - LABOGEO/UFPE (2016).

2.2 Physiographic aspects of the study area

2.2.1 Weather

The Rio Formoso estuary is located in the so-called Zona da Mata, whose characteristic climate is tropical humid, which according to the Kõppen classification is type As' - tropical, with early winter rains in autumn (with high annual rainfall totals of 1,800 to 2,000 mm, characterising a humid region with a strong rainfall gradient (CONDEPE/FIDEM, 2006). Rainfall is mainly caused by cyclones from the Atlantic Polar Front. These arrive on the northeastern coast with greater force during the autumn-winter period (SETUR/CPRH, 2011). In compensation for this high rainfall, solar radiation *is* intense, typical of the tropical belt, whose sea breeze contributes to a high evaporation rate.

The average temperature hovers around 26°C, with a small annual temperature range of around 4°C. The relative humidity in the rainy months is around 80 per cent, as a result of the low latitude, the proximity of the Atlantic Ocean and the air masses acting in the region.

In the region studied, the predominant wind direction is from the E quadrant - normal circulation - coming from the area of subtropical high pressure, i.e. the South Atlantic anticyclone, known as the trade winds. This mass has an upper thermal inversion with two layers: the lower, with lower and wetter temperatures, and the upper, with higher and drier temperatures. It has a stable character that ends with the arrival of disturbed currents (NIMER, 1977).

2.2.2 Vegetation

Originally, the Atlantic rainforest, with its exuberance and heterogeneity, covered the entire coastal strip of the state. Today, a few remnants of this primary vegetation are thought to be sparsely distributed throughout the southern coast of Pernambuco (CPRH, 2001). According to the Ministry of the Environment, the Rio Formoso Estuary's predominant vegetation cover is classified as pioneer formation formed by mangroves (pioneer formation with fluvial and/or lacustrine herbaceous influence), restingas (pioneer formation with arboreal marine influence) and coconut palms (secondary vegetation)

(CONDEPE/FIDEM, 2006) (figure 3). The mangroves are mainly represented by *Rhizophora mangle* (red mangrove), *Laguncularia racemosa* (black mangrove), *Avvicennia schaueriana* (white mangrove) and *Conocarpus erectus* (button mangrove). The restingas have a predominantly herbaceous-shrub formation, but there are also some trees, interspersed with shrubs and herbs (CPRH, 1998).

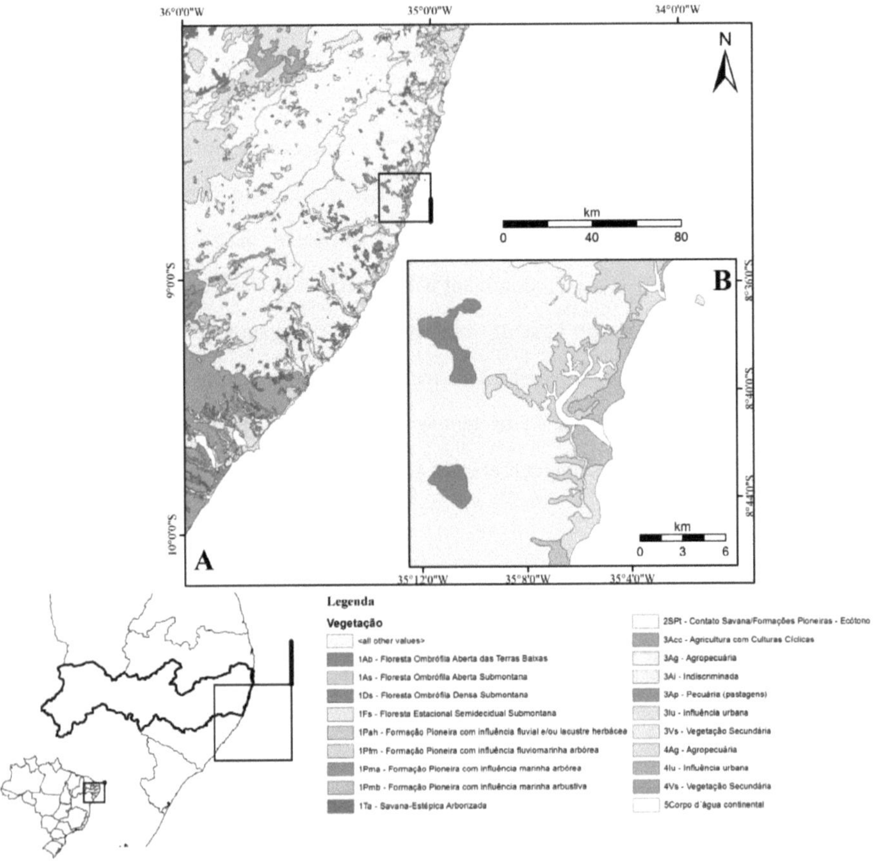

Figure 3: Vegetation map of the study area. Source: Diego Xavier - LABOGEO/UFPE (2016).

2.2.3 Water resources

The Rio Formoso estuarine system *is the* most representative water body within the Guadalupe APA (SETUR/CPRH, 2011). It is part of the group of small coastal river basins - GL 4 in the state of Pernambuco (CONDEPE/FIDEM, 2006), a region with high rainfall, making the fluvial regime of the watercourses perennial. According to Lira et al. (1979), the Formoso River estuary has a residence time of approximately 11-12 days. Although perennial, the Formoso River has low flows of approximately 4 m^3/s (COMPESA, 2006) (4.78 and 5.12 m^3/s (November and July, respectively) (LIRA, et al., 1979), which are intensified during the rainiest periods. The Formoso River rises in the north-western part of the municipality of the same name, on land belonging to the Vermelho Mill, where the headwaters of its two sources are located - the Vermelho and Serra d'Agua streams - which converge upstream of the Changuazinho Mill headquarters. From this confluence, now known as the Formoso River, it heads southeast, passing through the town of the same name. Three kilometres downstream, the Formoso River reaches the coastal plain, dominated by its wide estuary associated with the coastal tributaries made up of the Passos, Porto das Pedras and Lemenho rivers, whose open channels determine the formation of mangroves and salt marshes at the foot of the Cretaceous and Tertiary hills and hills, up to the recent marine terraces. Near its mouth, the Rio Formoso is joined on its right bank by the Rio Ariquindá, an important component of its basin (SETUR/CPRH, 2011).

2.2.3.1 Classification of the Rio Formoso estuary

Silva (2008) classified the Rio Formoso according to its hydrodynamics as being dominated by tides. In terms of geomorphology, the author classified it as being of tectonic origin, due to the geological faults and fractures that dominate the region. With regard to salinity, the estuary was classified as vertically homogeneous and well mixed. From a geological point of view, Lira et al. (1979) classified the estuary as a coastal plain due to the following aspects: presence of a central channel with a cross-section showing a slightly triangular trough; maximum depth of less than 15m; width to depth ratio greater than Im.

2.2.3.2 Division of the Rio Formoso estuary

The Rio Formoso estuary was divided by Lira et al (1979) into three distinct morphological zones:

- Upper estuarine zone: this corresponds to the area upstream of the largest island (Neri) in the estuary up to the maximum limit of penetration of the saline tide. The morphological characteristic is the presence of a channel with a depth of 2 metres or more and the existence of sandy-silt banks covered by mangrove vegetation;

- Middle estuarine zone: zone situated between the mouth of the Ariquindá River and the largest island in the estuary. The most significant morphological feature of this zone is a channel with an average depth of 7.05 metres, which is the axis of the ebb and flow of the waters and is practically all anchored on the right bank of the estuary;
- Lower estuarine zone: the area between the tip of Guadalupe and the mouth of the Arinquidá river. This is the widest part of the estuary, where there are two channels, one flowing and the other ebbing, located on the banks (no mangroves) and divided by a small sandy bank.

With regard to the Rio Formoso channel, Amaral (1992) divided it into three sectors: the upper fluvial channel (which runs from the vicinity of the town of Rio Formoso to its mouth in the bed of the Rio dos Passos), the central fluvial channel (wider and deeper than the first, corresponding to the channels of the Rio dos Passos and Ariquindá) and the mouth (which corresponds to the easternmost sector of the main channel).

In a more recent study, Silva (2008) divided the Rio Formoso estuarine system into three sectors:

- The upper sector, which includes the Lemenho, Passos and Porto das Pedras tributaries, has an average bathymetric level of Im. It should be noted that on the left bank of the Lemenho river there is a depression that reaches a depth of 9 metres (figures 4 and 5).

Figure 4: Upper estuarine zone with a view to the west of the shrimp ponds. Photograph by Roberto Barcellos, taken on 14/06/2011.

Figure 5: Upper estuarine zone with a view to the south of the shrimp ponds. Photograph by Roberto Barcellos, taken on 14/06/2011.

- Middle sector, stretch between the confluences of the Formoso and Ariquindá rivers, with a depth of around 3 metres. Its main feature is a channel downstream of the confluence of the Formoso River, positioned on the right bank of the estuary. This channel runs perpendicular to the confluence with the Ariquindá. Although the average depth is 3 metres, there are depths of more than 5 metres, which can reach up to 9 metres in stretches along this sector (figures 6 and 7).

Figure 6: View of the upper estuarine zone from the middle estuary sector. Photograph by Roberto Barcellos, taken on 14/06/2011.

Figure 7: Another view of the lower estuarine zone from the middle estuary sector. Photograph by Roberto Barcellos, taken on 14/06/2011.

- Lower sector, from the confluence with the Ariquindá River to the mouth, with an average bathymetre of 4m. There are two deeper sections in this sector with depths greater than 5m. One is located on the left bank of the estuary and the other is near the mouth, parallel to the reef line.

- The upper sector shows facies characteristics of low energy conditions (muddy sand, sandy mud and mud facies), associated with the shallow depth of the stretch, which favours the deposition of finer sediments. Slightly coarser sediments with gravelly sand and sand facies can be seen in the middle and lower sectors, consistent with the higher energy conditions due to the greater depth compared to the upper sector (figures 8, 9 and 10).

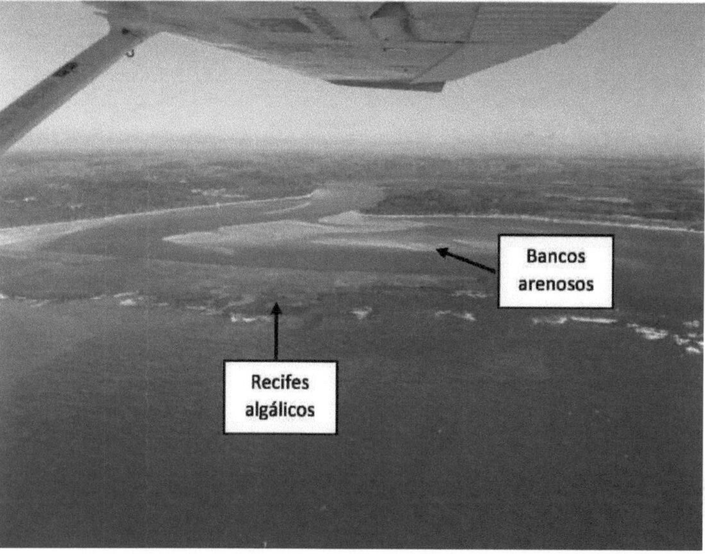

Figure 8: View of the mouth of the Rio Formoso estuary with the presence of sandy banks and algal reefs. Photograph by Roberto Barcellos, taken on 14/06/2011.

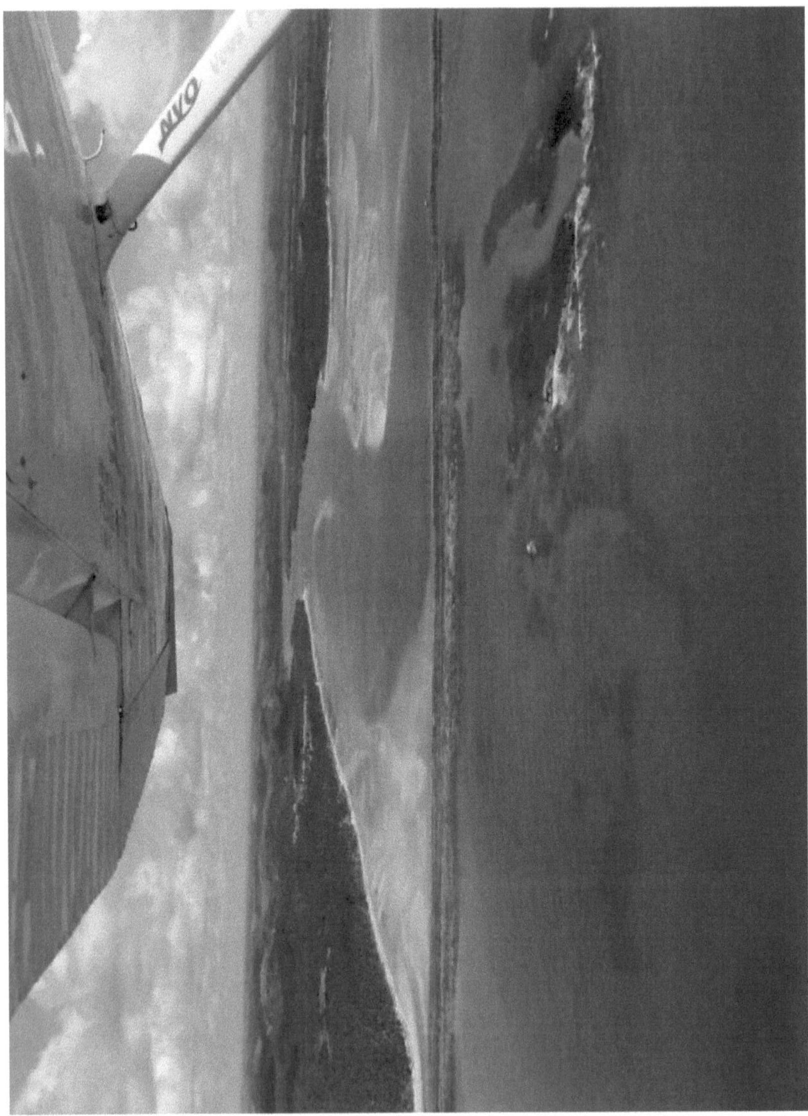

Figure 9: View of the mouth of the Rio Formoso estuary with the presence of sandy banks and algal reefs. Photograph by Roberto Barcellos, taken on 14/06/2011.

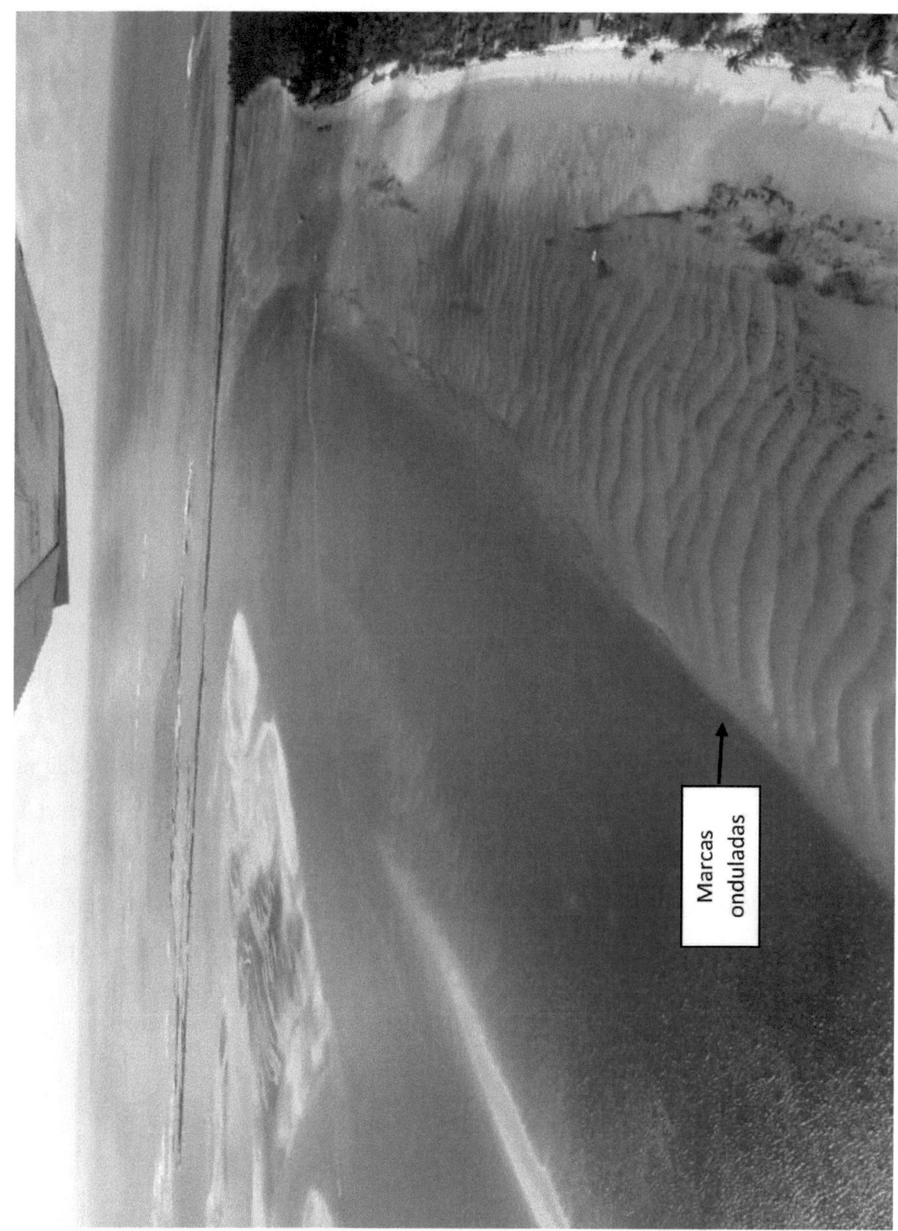

Figure 10: SE view of the mouth of the Rio Formoso estuary with the presence of sandy banks and undulating marks. Photograph by Roberto Barcellos, taken on 14/06/2011.

2.2.3.3 Bathymetry of the Rio Formoso estuary

Silva (2008) carried out a bathymetry and imaging survey of the area under study and observed that the bottom morphology of the estuary had positive and negative vertical relief. The positive vertical relief was attributed to the sandy banks and the negative vertical relief to the depressions and the deepest channel (figure 11). In the channel, bathymetric levels of less than 4m predominated, corresponding to the stretches of gentle slopes (upstream) and the deepest portions of the fluvial-estuarine channel had bathymetric levels of more than 5m, which could reach up to llm.

2.2.3.4 Physico-chemical water parameters

With regard to the physico-chemical parameters of the water, Silva (2009), in March 2008, measured some parameters in the surface water, such as: temperature, pH, Eh, electrical conductivity and salinity. The water temperature varied between 28.0 and 32.9° C, with the increase in temperature being related to the intensity of solar radiation and the variation in tides (ebb tides, lower temperatures, and flood tides, higher temperatures). The author also recorded pH data ranging from 6.50 to 7.80. Similar values, ranging from 6.67 to 8.30 obtained between the months of October (2005) and August (2006) had been reported for the estuary under study (PAIVA et al., 2008). Eh varied between 151 and 227 mV (oxidising conditions). Electrical conductivity showed values between 3.2 and 17.1 mS.cnT^{1}, which is common in estuarine environments where salinity is high and varies according to the tidal regime. The calculated salinity values varied between 7.5 and 9.2%o. In general, the lowest salinity values were found on the ebb tide and the highest salinity values on the flood tide.

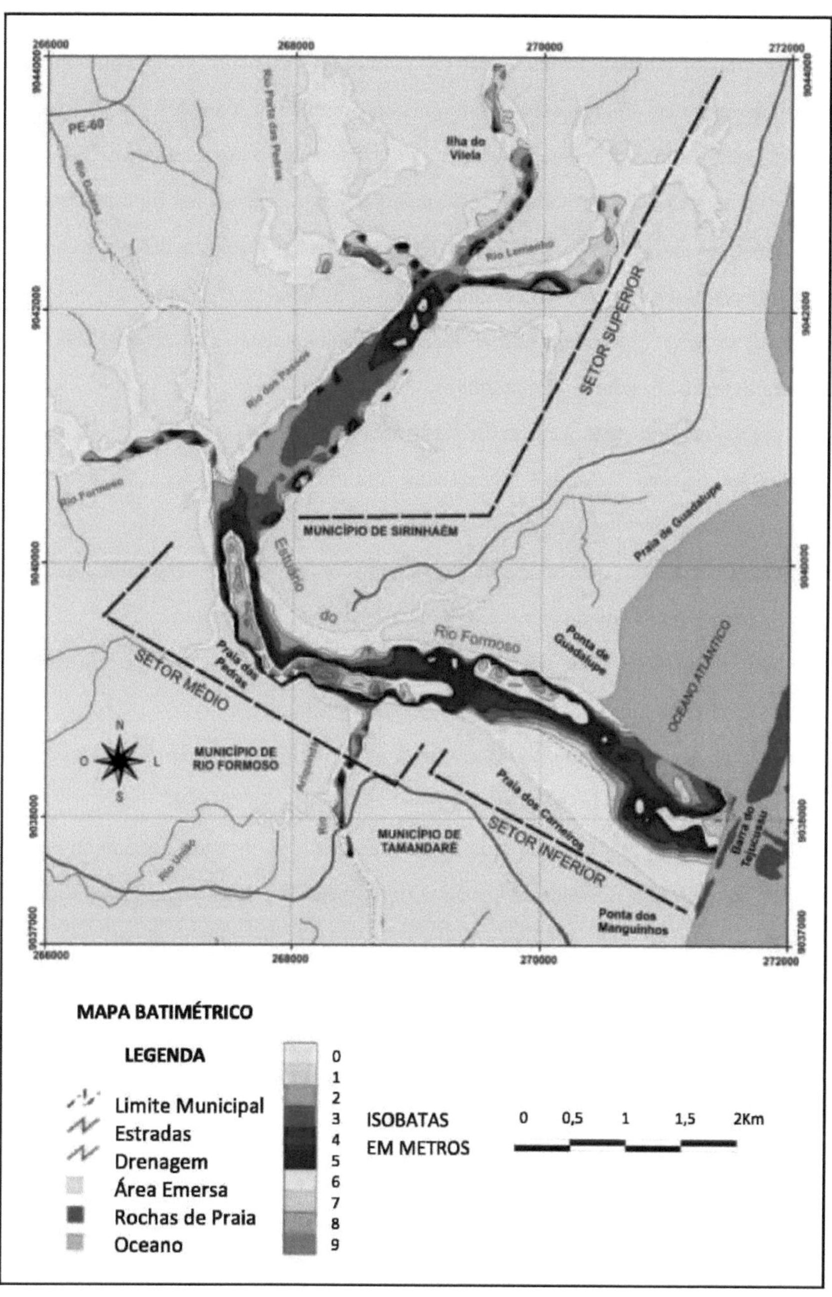

Figure 11: Bathymetric map of the Formoso River estuary area (Source: Silva, 2008).

2.3 Geological aspects of the study area

The geological substrate of the area is represented by two major geological domains: the crystalline basement represented by a gneissic-migmatitic complex of granitic to granodioritic composition of Palaeoproterozoic age (Pemambuco-Alagoas Massif) and the sedimentary sequence of the Pernambuco Sedimentary Basin represented by the Cabo and Barreiras formations and Quaternary deposits (figure 12). Partially capping these units are the Tertiary sediments of the Barreiras Formation, followed by Quaternary sediments, including ancient marine terraces (Pleistocene and Holocene), palaeomangues, beach rocks and more recent coastal deposits: beaches and mangroves.

The geological units that occur in the study area are:

- Crystalline basement, formed by igneous and metamorphic rocks of granitic to granodioritic composition, with important outcrops in the region;
- The Cabo Formation, of Cretaceous age, sits discordantly on the crystalline bedrock and *is* fundamentally made up of conglomerate, arkose and claystone;
- Barreiras Formation, made up of thick sandstones interspersed with conglomeratic, clayey levels rich in iron oxide (LIMA FILHO, 1998);
- Pleistocene terraces, essentially sandy, with an altitude of between 3-9m, devoid of mollusc shells (Quaternary deposits);
- Holocene terraces are laid out parallel to the coastline, in wide continuous strips, formed by white unconsolidated sands, with the presence of mollusc shells, reaching heights of up to 3m and can be seen along the coastal plain (alluvial deposits);
- River sediments found in the areas of river valleys and troughs. They are represented predominantly by sands and silts;
- Lagoon deposits are found in shallow, elongated depressions made up of muddy sediments with organic matter;
- Mangrove deposits are found in the innermost part of the estuary in areas upstream of rivers, in smaller tributaries and in tidal channels. They are rich in organic matter and are essentially made up of silt and clay;
- Coastal deposits found on the beach line consisting of well-sorted quartz sands.

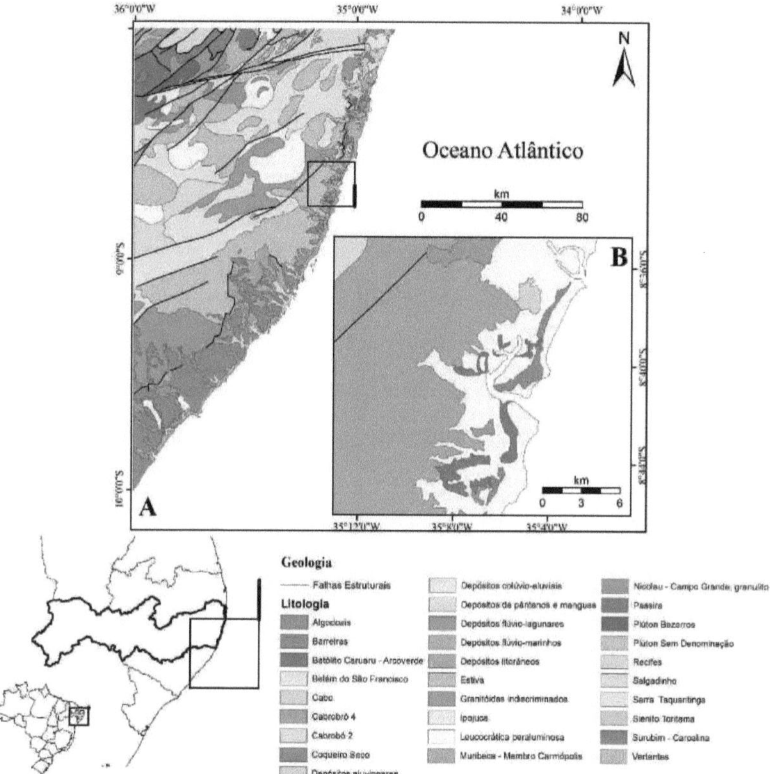

Figure 12: Geological map of the study area. Source: Diego Xavier - LABOGEO/UFPE (2016).

1.1.1 Channel Sedimentology

Silva (2008) carried out a study to find out the textural distribution of the sediments that line the estuary's main and secondary channels. He collected 220 samples in transverse profiles of the channels at average distances of 200 metres.

In each profile, he tried to collect samples from the banks and bed of the channels. As a result, he found that the distribution of the gravel fraction played little part in the total of the samples analysed, finding the highest concentrations in the samples from greater depth and near Praia das Pedras, on the right bank of the middle estuary channel. Samples with sand percentages above 90 per cent predominated. The highest proportion of mud was found upstream of the estuary, and the lowest in the middle and lower stretches of the system. As for the distribution of sediments according to average diameter, Silva (2008) found a predominance of medium and fine sands.

2.4 Gemorphological aspects

In the study area, the crystalline domain, coastal tablelands, collinear modelling and the coastal plain make up the landscape forms (CPRH, 1998). The crystalline domain is characterised by hilly terrain, also known as "mares de morros", with irregular morphology and altimetry reaching just over lOOm. The coastal tablelands are characterised by sediments from the Barreiras Formation with altimetry between 40 and lOOm. The hills are characterised by the presence of small gentle hills with altimetry between 10 and 40m. The estuary itself is part of the coastal plain in which river terraces, tidal flats and Pleistocene and Holocene marine terraces can be identified (figure 13).

River terraces are the product of erosion and deposition associated with fluvial transport in the pre-quatemary units of this system. Tidal flats are found in areas with almost zero slope gradients. They are favourable environments for fluvial-marine sedimentation processes. The marine terraces, on the other hand, are characterised by features created during the oscillations of the sea during the Quaternary, which have been subdivided into two types according to their age of deposition: Pleistocene and Holocene.

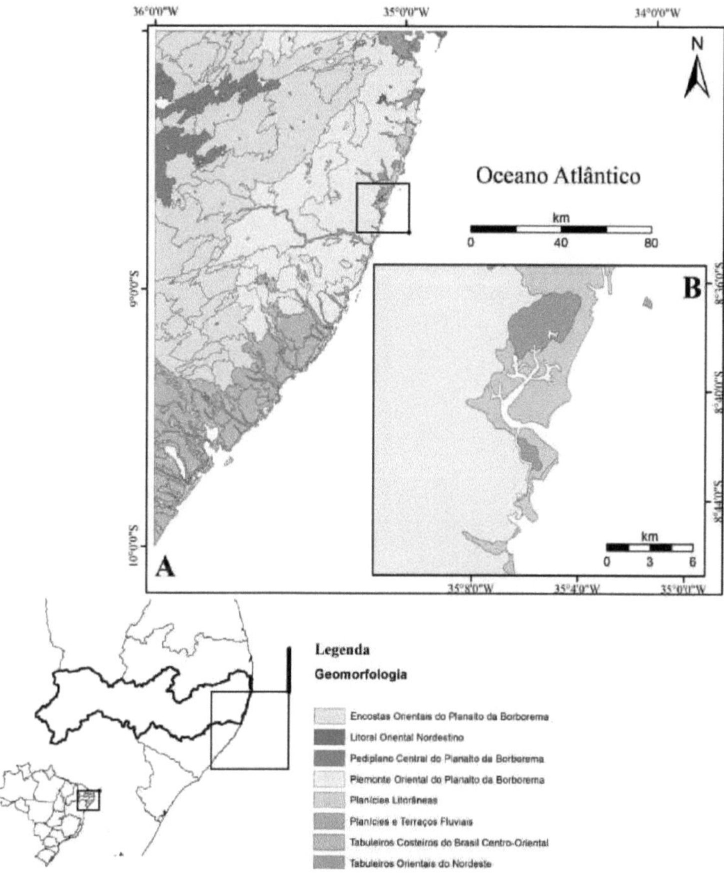

Figure 13: Geomorphological map of the study area. Source: Diego Xavier - LABOGEO/UFPE (2016).

2.5 Land use and occupation

The land use and occupation units shown in figure 14 express the main activities carried out in the area. The following have been identified: mangrove areas around the estuary; coconut cultivation on the coast, around the estuary, occupying the marine terraces and tablelands, being replaced by holiday allotments and hotel structures; urban sprawl, mainly in the southernmost region of the estuary; areas of shrimp ponds around the estuary in the upstream region; and extensive agricultural land with polycultures, including sugar cane cultivation.

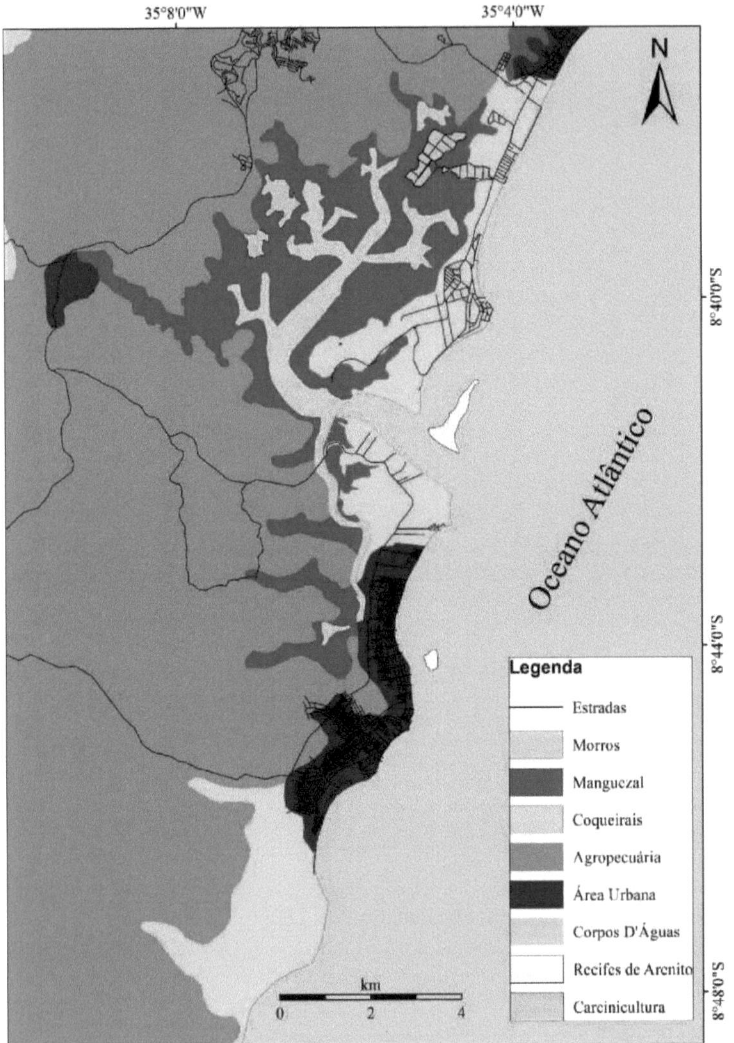

Land use and occupation map for the Rio Formoso estuary region. Source: Diego Xavier - LABOGEO/UFPE (2016).

CHAPTER 3

Seasonal sedimentology of the Rio Formoso estuary

3.1 Shepard's (1954) classification of easy textures

The distribution of sediments, based on Shepard's (1954) triangular diagram, for the rainy season/2014 and summer/2015 (figures 15 and 16) indicated a great similarity for the two periods studied, with a predominance of sandy sediments, mainly sands (77.08% of the samples for both periods). In addition, according to Shepard's classification (1954), the study area had three distinct sediment textural facies for the rainy season/2014 and four distinct sediment textural facies for the dry season/2015 (tables 2 and 3, respectively).

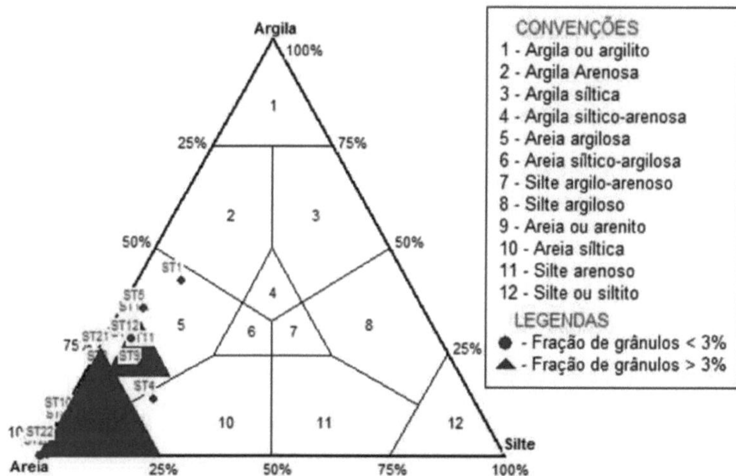

Figure 15: Shepard's triangular diagram (1954) for the rainy season (September 2014).

Table 2: Shepard's (1954) easy textural classification for the rainy season (November 2014).

Easy to texture	Percentage (%)
Clayey sand	37,50
Sand	58,33
Silica sand	4,17

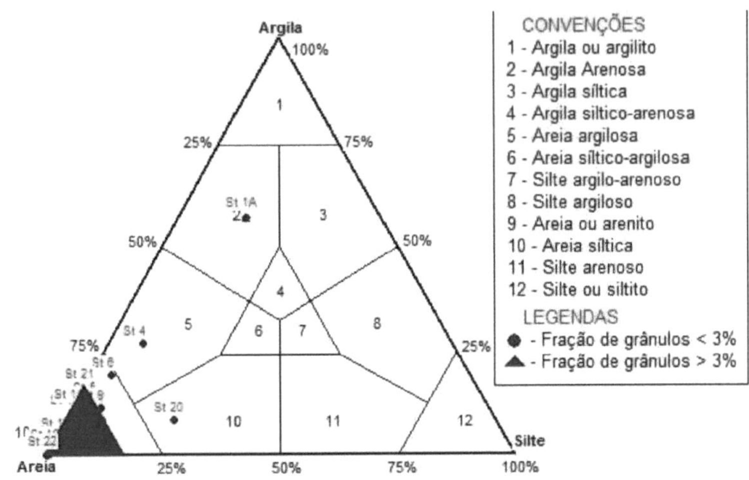

Figure 16: Shepard's triangular diagram (1954) for the dry period (March/2015)

Table 3: Shepard's (1954) easy textural classification for the dry period (March/2015).

Easy to texture	Percentage (%)
Sandy clay	4,17
Clayey sand	4,17
Sand	87,49
Silica sand	4,17

3.2 Mean diameter (Folk & Ward, 1957)

The maps of sediment distribution according to average diameter (figures 17 and 18) showed great similarity between the wet period/2014 and the dry period/2015. Both periods showed heterogeneous results, with sediments ranging from medium silt to coarse sand for the wet period/2014 and from fine silt to coarse sand for the dry period/2015. With a tendency for finer sediments to be concentrated in the middle and upper estuary, corresponding to the meandering and more confined portion of the Rio dos Passos Channels, associated with the mouths of the small tributaries, Rio Lemenho and Porto das Pedras. Despite this general pattern, a much greater predominance of sand was observed in the dry/2015 period (83.33 per cent) than the rainy period (70.83 per cent).

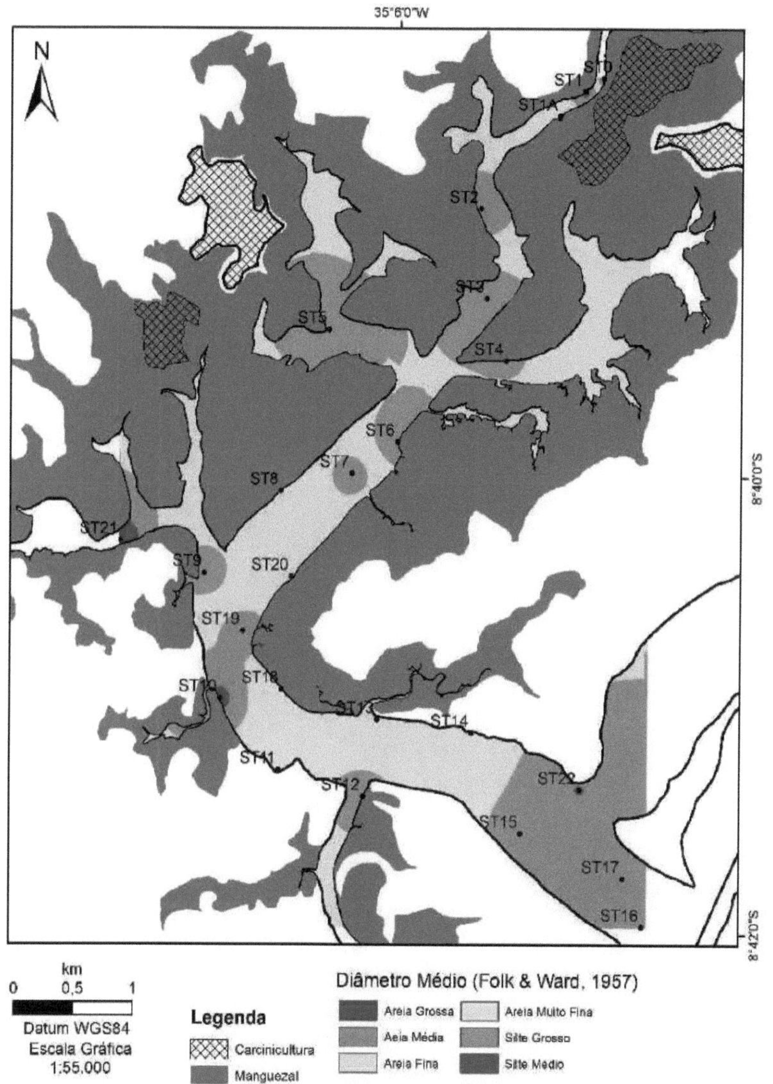

Figure 17: Average sediment diameter during the rainy season (September 2014). Source: Diego Xavier- LABOGEO/UFPE (2016).

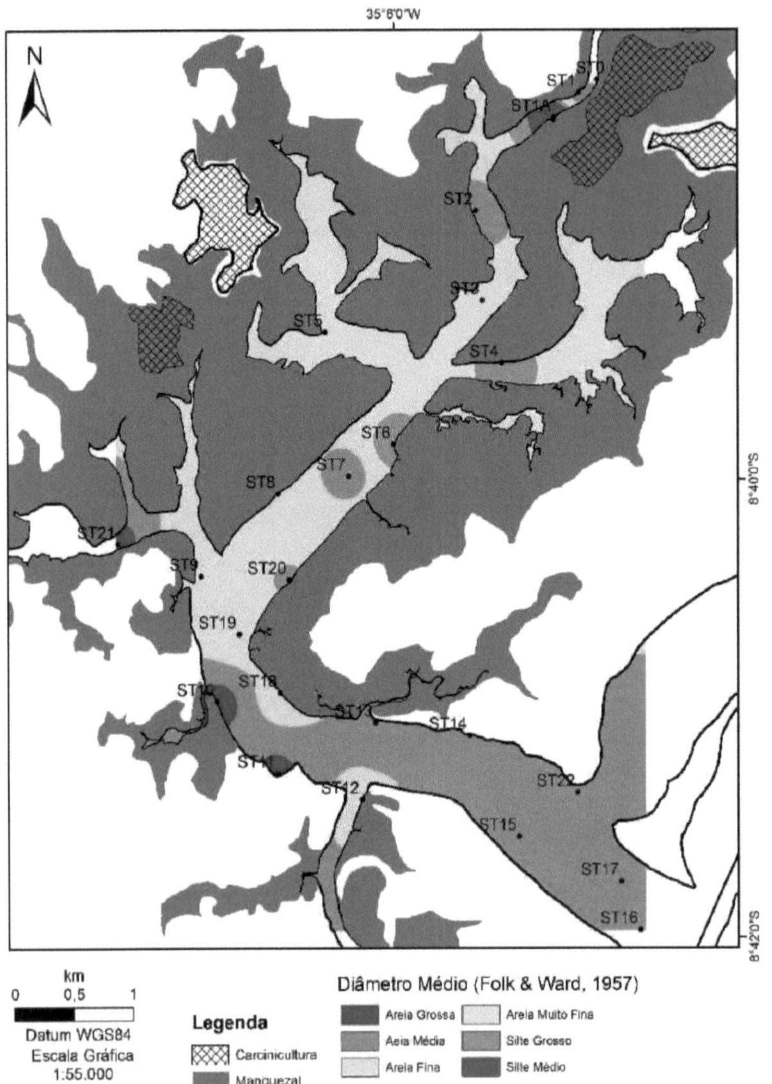

Figura 18: Average sediment diameter during the dry period (March/2015). Source: Diego Xavier - LABOGEO/UFPE (2016).

18.3 Degree of selection (Folk & Ward, 1957)

Regarding the degree of sorting (figures 19 and 20), there is a predominance, regardless of the season, of poorly to very poorly sorted sediments, in around 60% of the samples (66.67% in the rainy season/2014 and 58.33% in the dry season/2015). For both periods, the samples with the best degree of sorting are associated with sandy sediment deposits and in the vicinity of the mouth of the Rio Formoso estuary. There was an increase in the degree of selection in the dry period/2015 compared to the rainy period/2014, due to the greater number of sandy samples. And consequently a greater number of samples with the degree of selection considered to be well selected (samples ST18 and ST19).

18.4 Gravel, sand and mud fractions

The distribution of the gravel fraction is similar for the two periods studied (figures 21 and 22). It can be seen that in the distribution of gravel, sediments with contents below 1% predominate for both the rainy period/2014 and the dry period/2015 (75.0% and 62.5% of the samples in the sample mesh, respectively). It can also be seen that the sample with the most gravelly material in the rainy period/2014 (ST21), located at the mouth of the Formoso River, suffered a reduction in its percentage of the gravel fraction in the dry period/2015 (25.46% to 16.46%). There was also a general reduction in the percentage of the gravel fraction in the dry period/2015. The distribution of sands is also similar for the two periods studied (figures 21 and 22). Sediments with contents above 75% predominate for both periods (58.33% of the samples from the rainy season sample mesh/2014 and 87.50% of the samples from the dry season sample mesh/2015). It was observed that the cores of sandier material occur throughout the lower estuary, mainly from the middle sector towards the mouth. Patches of muddy material, i.e. with sand content below 50%, occur at some points in the upper estuary, such as ST 00 and ST 01 (rainy season/2014) and ST 01A (dry season/2015). It was also observed that the sand content increased throughout the estuary in the dry period/15.

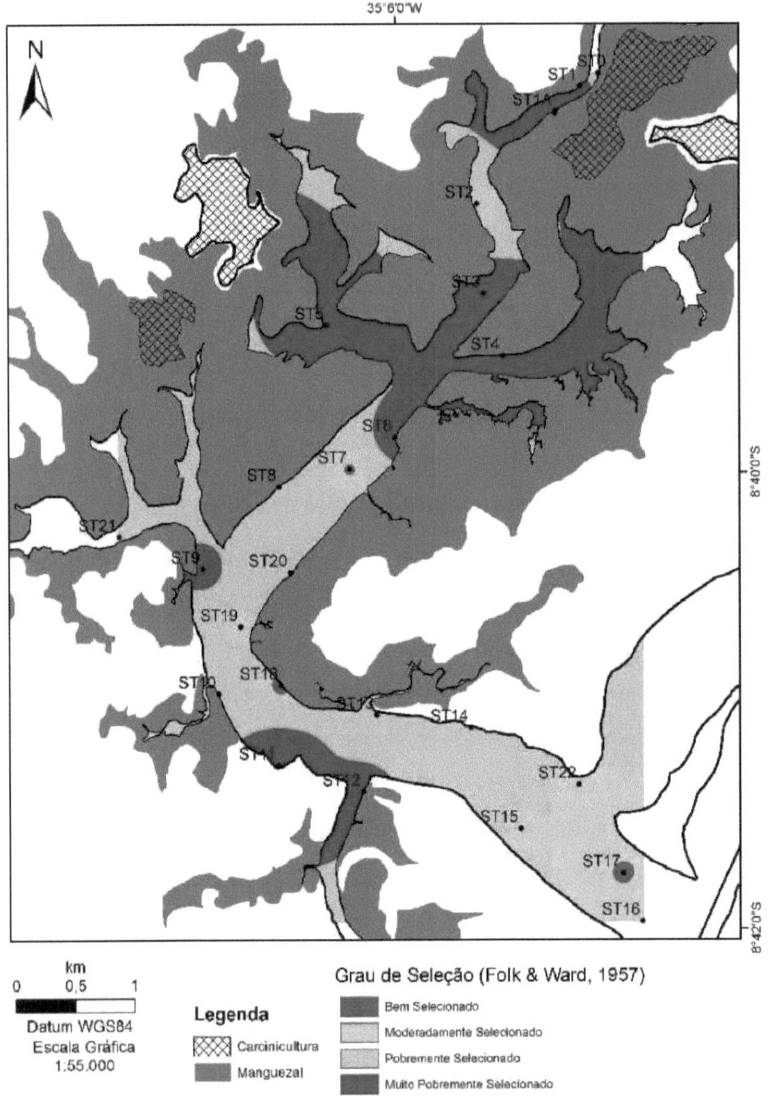

Figure 19: Degree of sediment selection during the rainy season (September 2014). Source: Diego Xavier- LABOGEO/UFPE (2016).

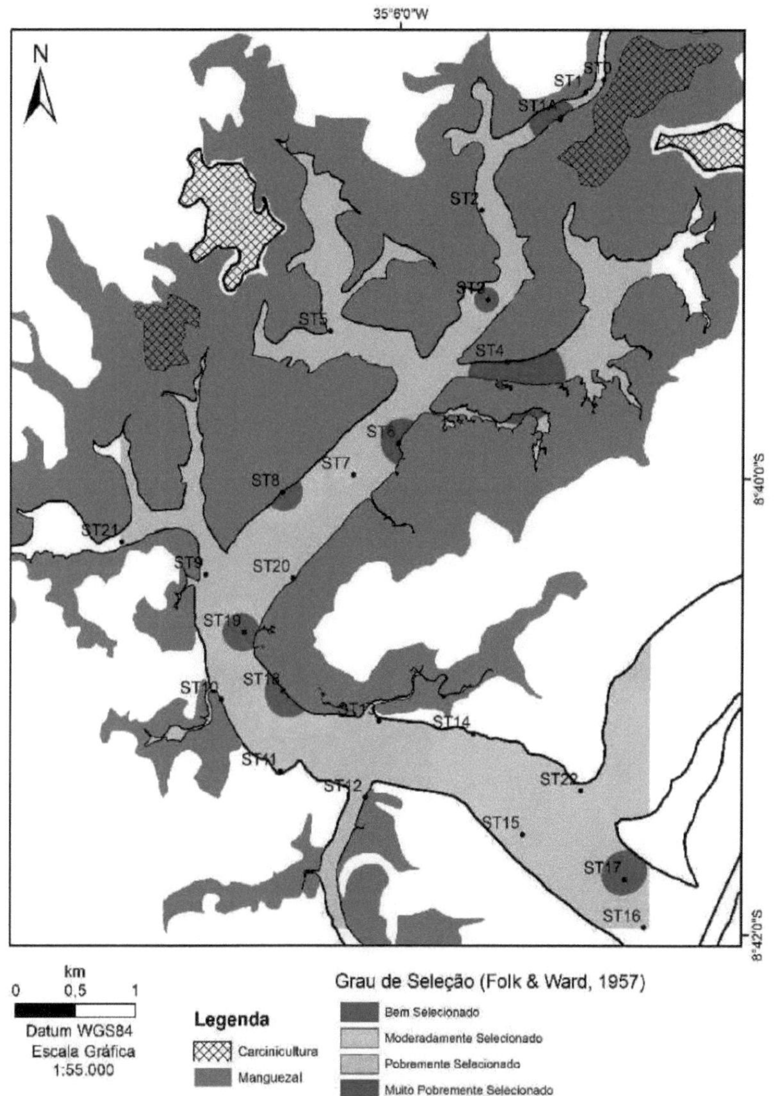

Figure 1: Degree of sediment selection in the dry period (March/2015). Source: Diego Xavier -LABOGEO/UFPE (2016).

Also during the dry season, there were patches of muddy material only in the area of station ST 1A (28.05% sand and 70.41% mud), located in the upper estuary.

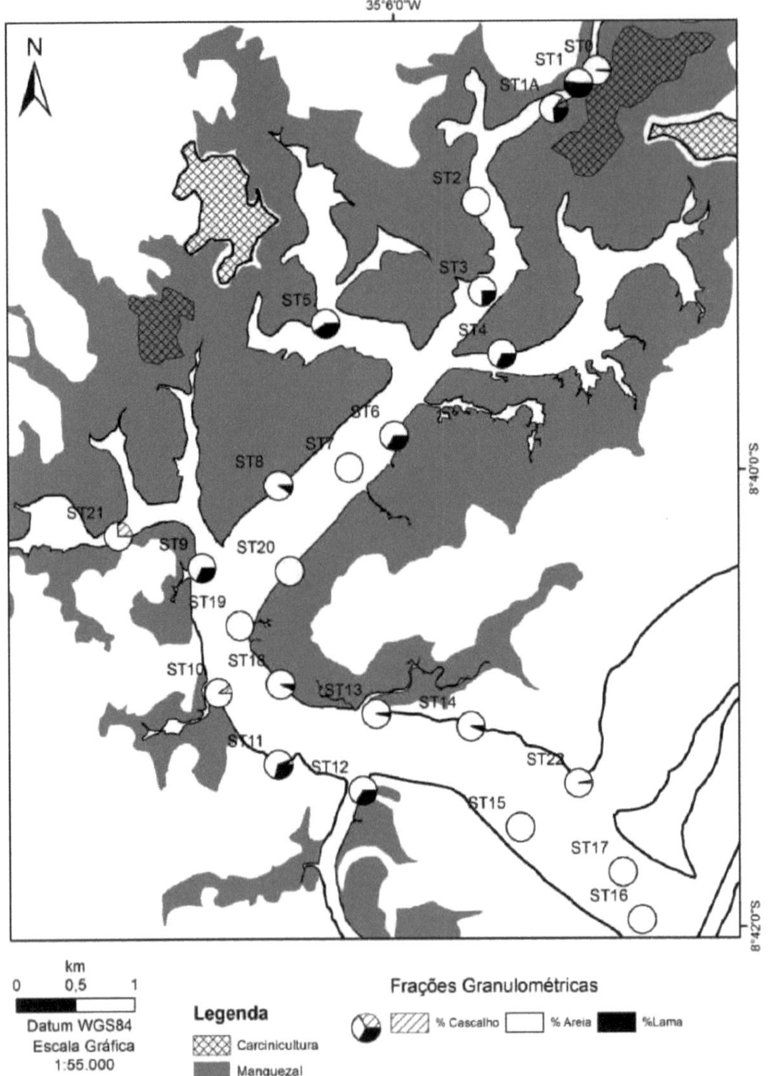

Figura 21: Gravel, sand and mud fractions (%) of sediments in the rainy season (September/2014).

Source: Diego Xavier - LABOGEO/UFPE (2016).

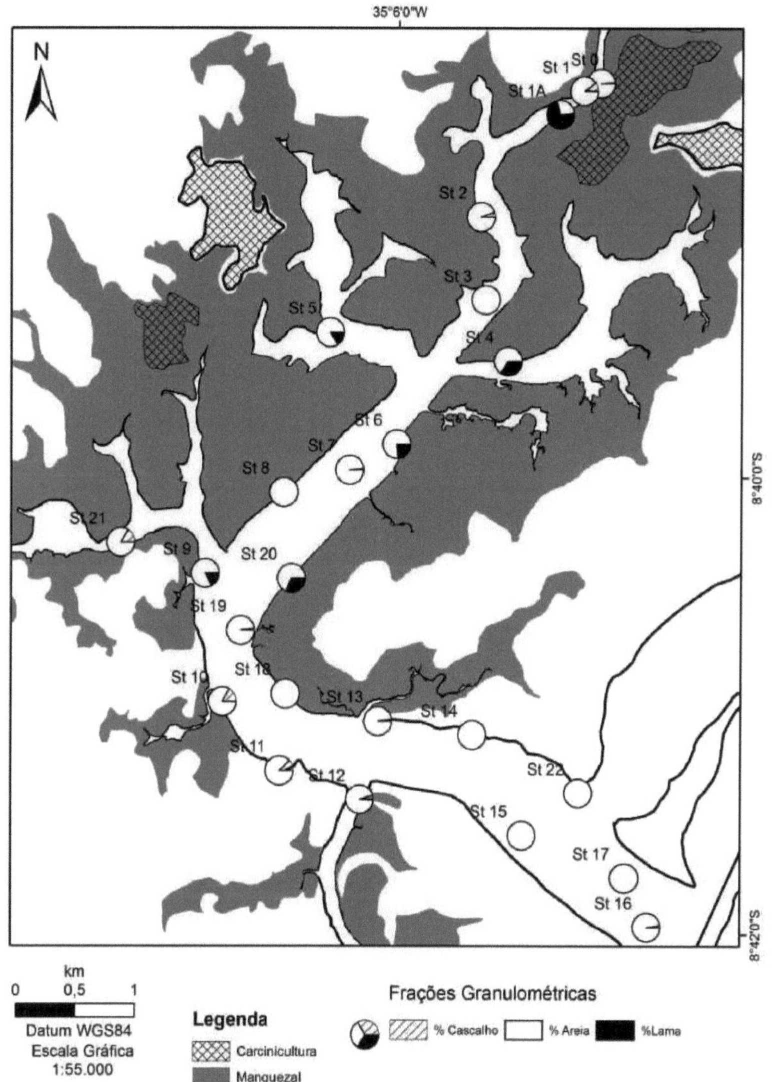

Figura 22: Gravel, sand and mud fractions (%) of sediments in the dry period (March/2015). Source: Diego Xavier - LABOGEO/UFPE (2016).

In the 2014 rainy season, the mud fractions showed a distribution that indicates a predominance of this fraction in the samples located in the upper estuary, further upstream. The sample with the highest percentage of mud (ST 01 51.73%), classified as medium silt according to Shepard's (1954) easy textural diagram, is located upstream of the estuary. Patches with contents of more than 25% can be seen almost everywhere in the estuary.

In the dry period/2015, the mud fractions showed a distribution that indicates a drop in fine sediment content in the samples from this period, including at the stations in the upper estuary. Samples with contents of less than 1% predominated, corresponding to 70.83% of the samples in the sample mesh for this period.

3.5 Compositional analysis of the sandy fraction

In the compositional analysis of the sandy fraction, the following were identified in the samples: coral fragments (CORAL), gastropods (GAST), foraminifera (FOR), scaphopod (ESC), echinoderm (EQUI), bryozoan (BRIO), bivalve (BIV), polychaete (POL), diatom (DIAT) and some biogenic fragments that were not identified (FBNI). There were also grains of quartz (QTZ), mica (MICA), plant fragments (FRAGV), heavy minerals (MPES), limonitised quartz (QTZL) and rock fragments (FRAGR). The acronym (TBM) was used for the total marine biogenic group and (TTR) for the terrigenous constituents.

In this study, the results obtained in the 0.500 mm (lo) and 0.250 mm (2©) fractions were interpreted differently. According to Pilkey et al. (1967, apud MAHIQUES et al., 1998), bioclastic fragments are more commonly found in the 0.500 mm (lo) fraction than in the 0.250 mm (2©) fraction.

In general, for both periods studied, there was a predominance of sediments with terrigenous influence of quartz composition (B.M. index = < -0.02), with B.M. indices predominating. < -0.92 (83.33% for the rainy season samples in both fractions and 58.33% for the dry season samples in both fractions), with the exception of sample 16 (the outermost station in the estuary), which showed a higher content of marine biogenic grains in the 0.500mm mesh for the rainy season (B.M. = 0.68) and in the 0.250mm mesh for the dry season (B.M = 0.6) (figures 23, 24, 25 and 26).

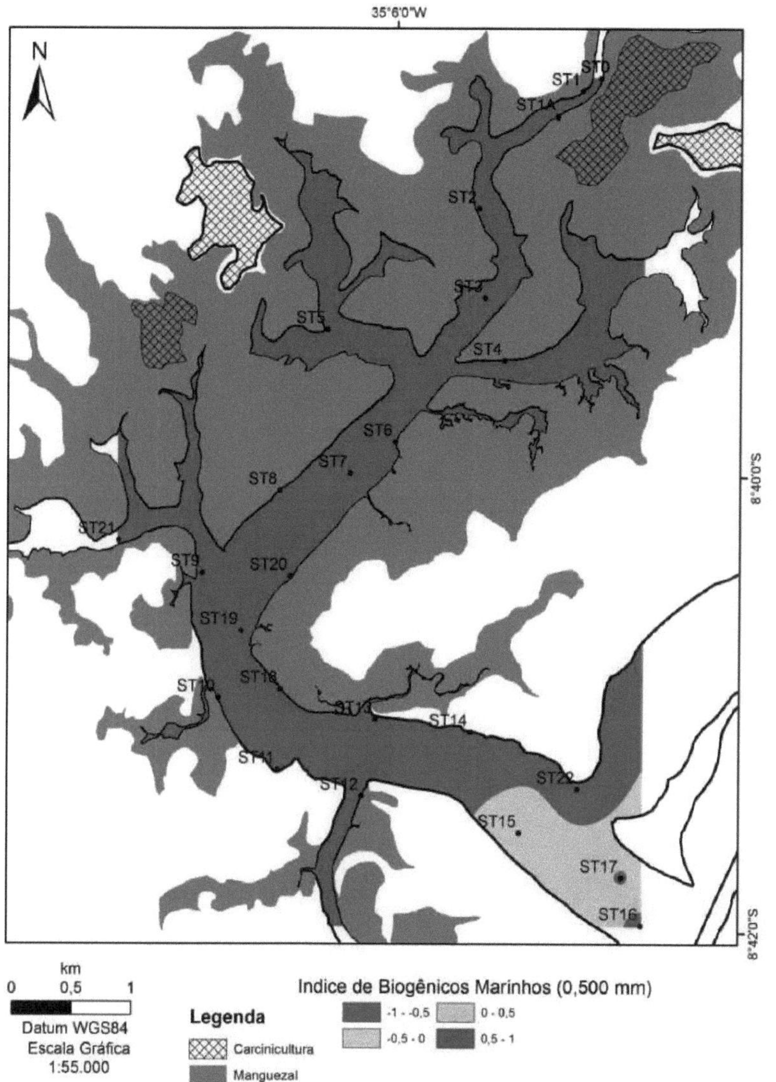

Figura 23: Map of sediment distribution according to the marine biogenic index (B.M) of the fraction (1©) in the rainy season (September/2014).

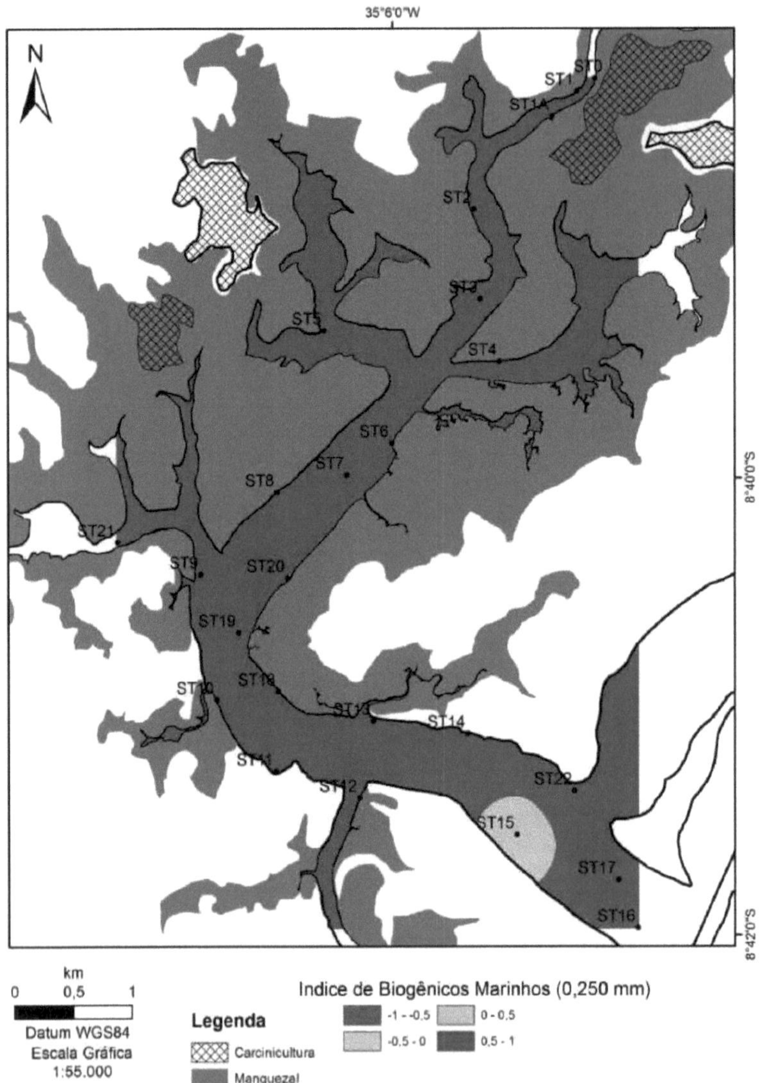

Figura 24: Map of sediment distribution according to the marine biogenic index (B.M) of the fraction 0.250mm (2<o) in the rainy season (September/2014).

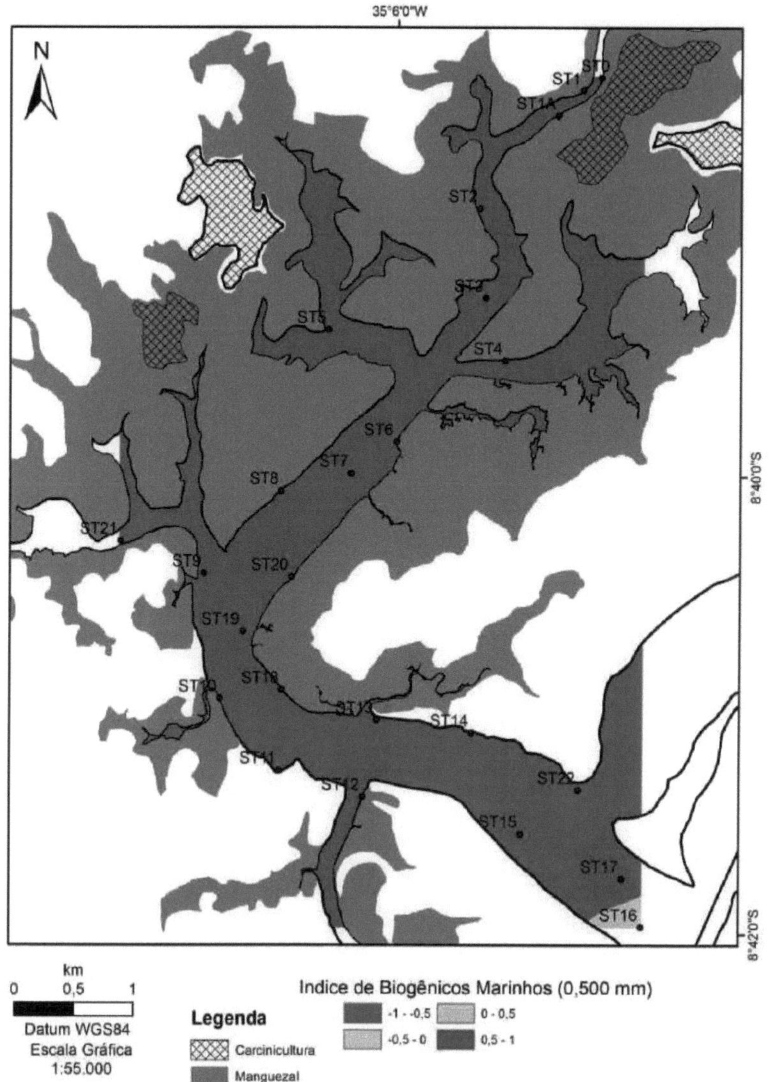

Figura 25: Map of sediment distribution according to the marine biogenic index (B.M) of the fraction (lo) in the Rio Formoso estuary for the dry period (March/2015).

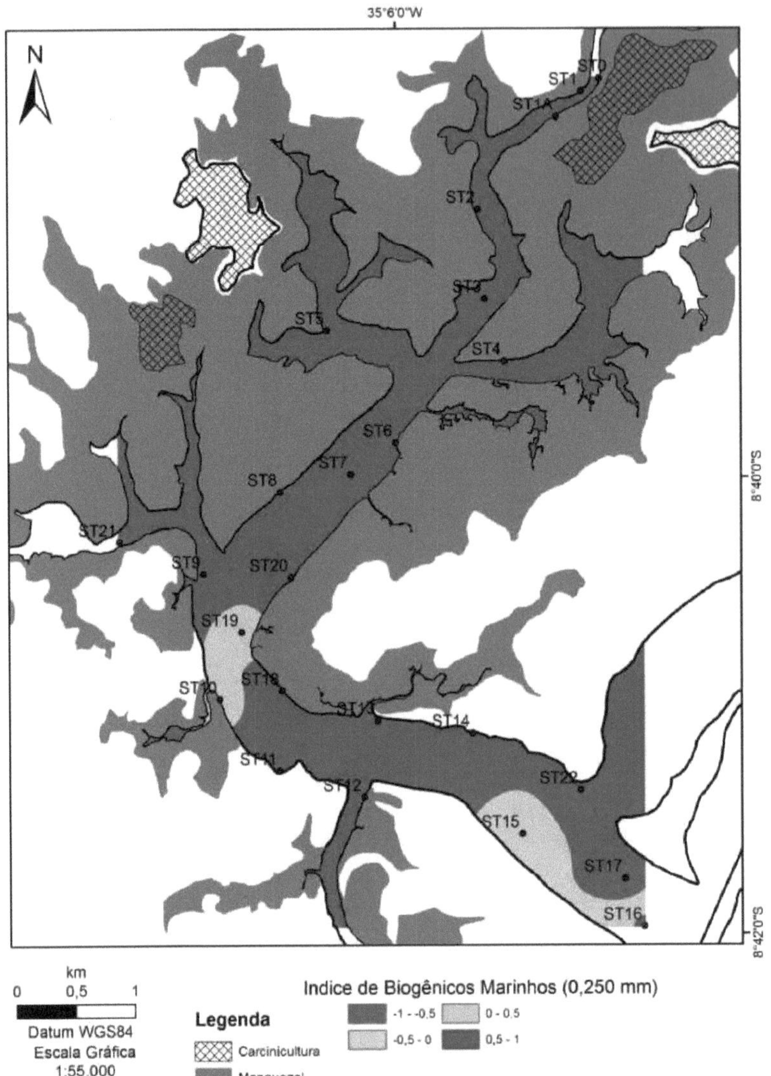

Figura 26: Map of sediment distribution according to the marine biogenic index (B.M) of the fraction (2co) in the Rio Formoso estuary for the dry period (March/2015).

3.6 Morphoscopic analysis of the grains

With regard to the surface texture of the sediment grains, a predominance of matte texture was observed for the two periods analysed (95.83% and 87.50% for the 0.500mm and 0.250mm sieves, respectively, for the samples from the rainy season and 91.16% and 79.16% for the 0.500mm and 0.250mm sieves, respectively, for the samples from the dry season).

With regard to the degree of roundness, there was a predominance of subangular grains in the finest fraction (0.250mm) of both periods (33.33% and 45.83%, respectively) and in the coarsest grain fraction (0.500mm) of the rainy period (54.16%), indicating a possible inefficient transport for both periods studied. As for the coarser grain fraction (0.500mm) from the dry period, there was a predominance of subrounded grains (41.66%). With regard to grain sphericity, high sphericity was predominant for the grain samples in both fractions analysed for the two periods studied. With a slight increase in the percentage of samples with low sphericity for the dry period in the 0.250mm mesh (table 4).

Table 2: Frequency of the morphometric characteristics of the sediments studied for the rainy period (RF1) and dry period (RF2) in percentages (%).

	RF1		RF2	
Sieves	0,250	0,500	0,250	0,500
Matte	87.50	95.83	75.00	66.67
Brilliant	12.50	4.17	20.83	12.50
Without Material	0.00	0.00	4.17	20.83
Angular	29.17	20.83	33.33	16.67
Subangular	33.33	54.17	45.83	20.83
Subrounded	29.17	25.00	16.67	41.67
High sphericity	79.17	83.33	62.50	95.83
Low sphericity	20.83	16.67	37.50	4.17

CHAPTER 4

Environmental geochemistry

4.1 Biodetrital carbonate content

The map of biodetrital carbonate contents drawn up for the samples from the 2014 rainy season (figure 27), whose values ranged from 0.20 to 57.40%, indicated the occurrence of lithoclastic sediments (91.67%), i.e. with contents below 30% CaCCb according to Larssoneur *et al.* (1982). Contents above 30% (ST 12, 43.6% and ST 16, 57.4%) are only found in the lower estuary region. The sample from ST 16 has the highest carbonate content and is located closest to the algal reef banks, representing the outermost station of the estuary.

The map of biodetrital carbonate contents for samples from the dry period/2015 (figure 28), whose values ranged from 1.00 to 45.30 per cent, also showed a predominance of lithoclastic sediments (95.83 per cent). Compared to the rainy season, there was a total reduction in total CaCOa content of 66.67% of the samples from the sample mesh. And a considerable increase for sample ST1A, located upstream of the estuary (1.9% to 17.0%). In general, for this period, the highest carbonate contents were found in the samples also collected in the lower estuary near the mouth of the stations (ST16, ST15, ST12) with contents of 45.30%, 21.50% and 16.50%, respectively.

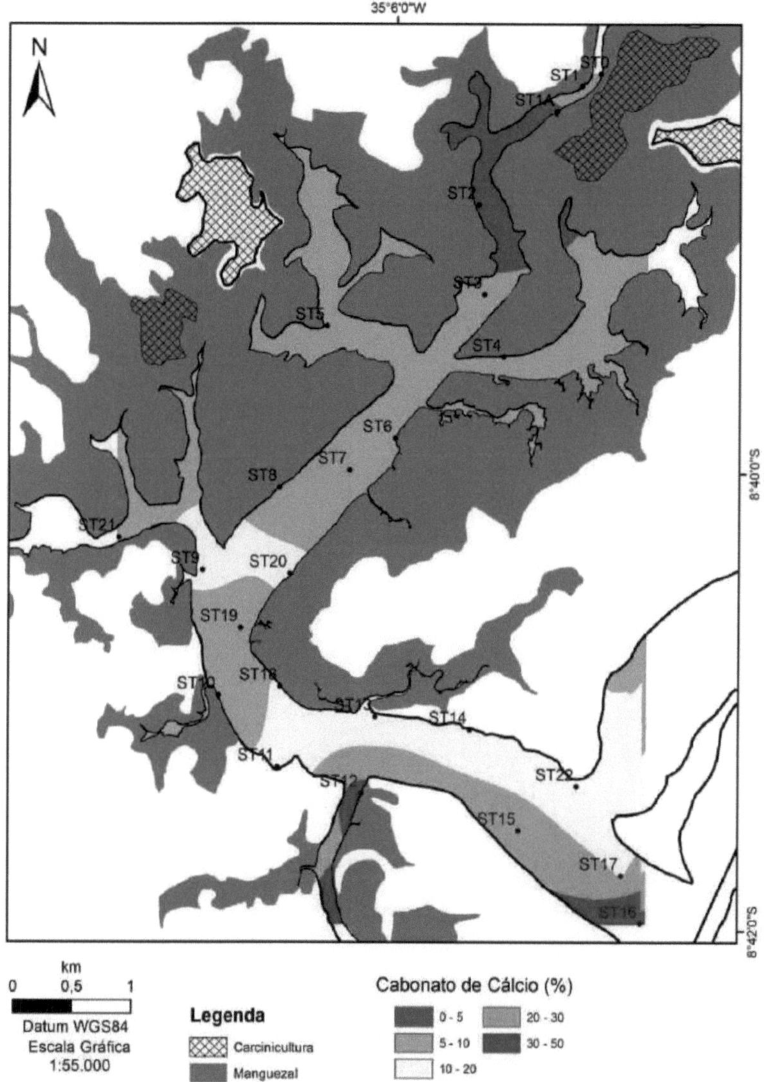

Figura 27: Calcium carbonate content (%) of sediments in the rainy season (September/2014). Source:
Diego Xavier - LABOGEO/UFPE (2016).

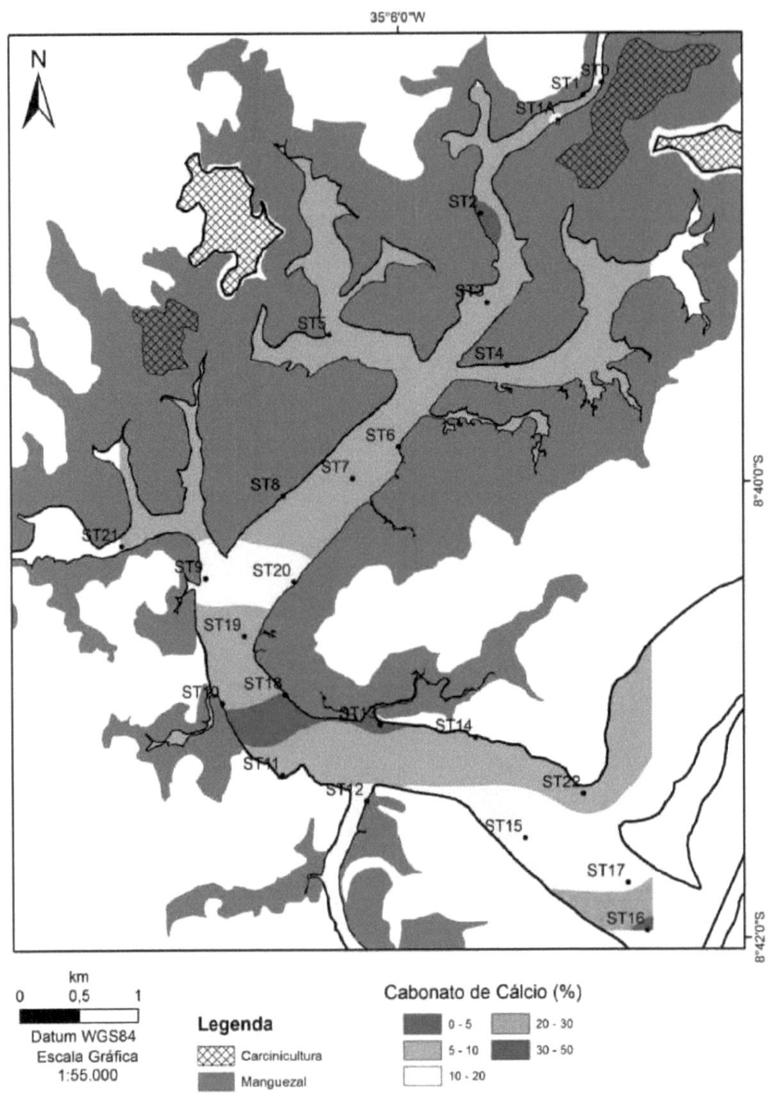

Figura 28: Calcium carbonate content (%) of sediments in the dry period (March/2015). Source: Diego Xavier - LABOGEO/UFPE (2016).

4.2 Total Organic Matter (MOT) content

The Total Organic Matter content (figure 29) in the 2014 rainy season ranged from

1.07% to 29.57%, with a predominance of values below 10% (70.83% of the samples) generally associated with sandier sediments located in the middle to lower estuary (mouth).

The percentages of Total Organic Matter in the dry period/2015 (figure 30) ranged from 0.80 to 35.77%, with a predominance of values also below 10% (75% of the samples) associated with sandier sediments also located in the middle to downstream region. During this period, the highest content was found in sample ST1A upstream of the estuary associated with a muddier sediment (fine silt). And although this content increased from the rainy period in 2014 to the dry period in 2015 (14.93% to 25.77%), overall 66.7% of the samples had a reduction in their MOT percentage.

4.3 Total organic carbon (TOC) content

The total organic carbon content shown in Figures 31 and 32 (rainy season/2014 and dry season/2015, respectively) varied between 0.11 and 7.00% (rainy season/2014) and between 0.11 and 6.01% (dry season/2015). The rainy season showed a predominance of samples with TOC contents between 0.0 and 2.0% (representing around 66.7%), while the dry season/2015 showed a predominance of samples with TOC contents between 2.0 and 4.0% (representing around 62.50%). But in general, most of the samples had low levels and were associated with sandy sediments, distributed throughout the estuary from the most upstream portion to the mouth. With the highest TOC content associated with sample ST1A for both periods studied, rainy/2014 and dry/2015 (7.00% and 6.01%, respectively), located in the upper estuary near the outflow of effluents from shrimp ponds.

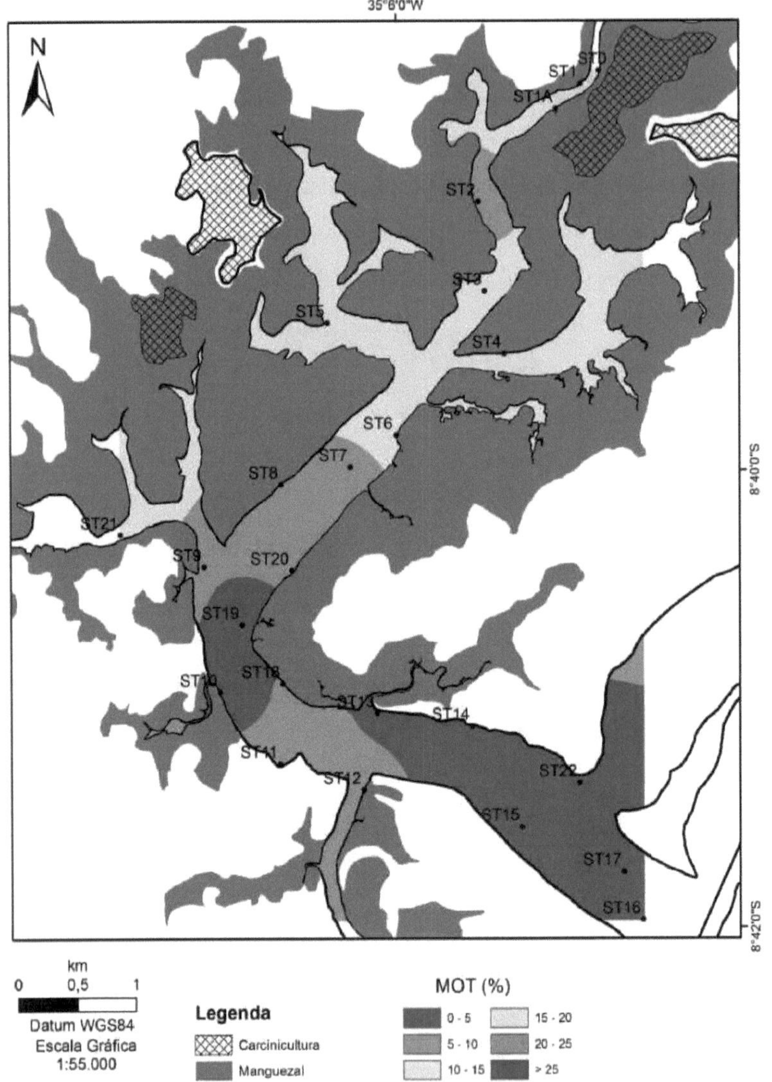

Figura 29: Total Organic Matter content (%) of sediments in the rainy season (September/2014). Source:
Diego Xavier - LABOGEO/UFPE (2016).

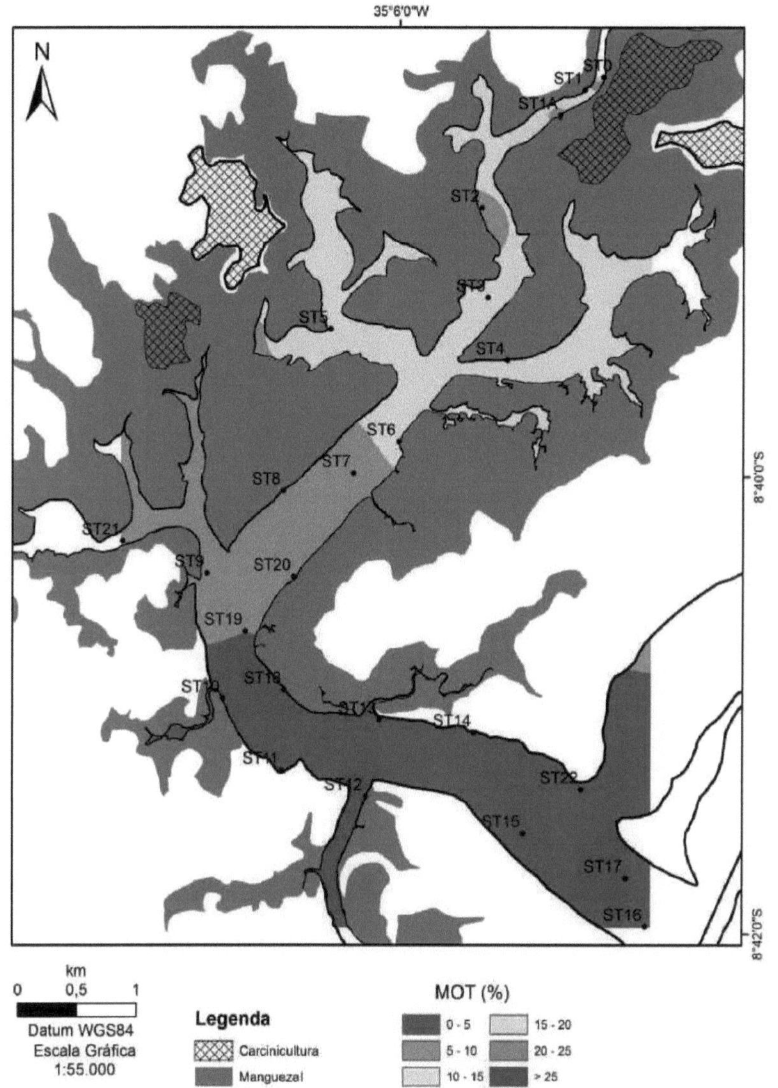

Figura 30: Total Organic Matter content (%) of sediments in the dry period (March/2015). Source: Diego Xavier - LABOGEO/UFPE (2016).

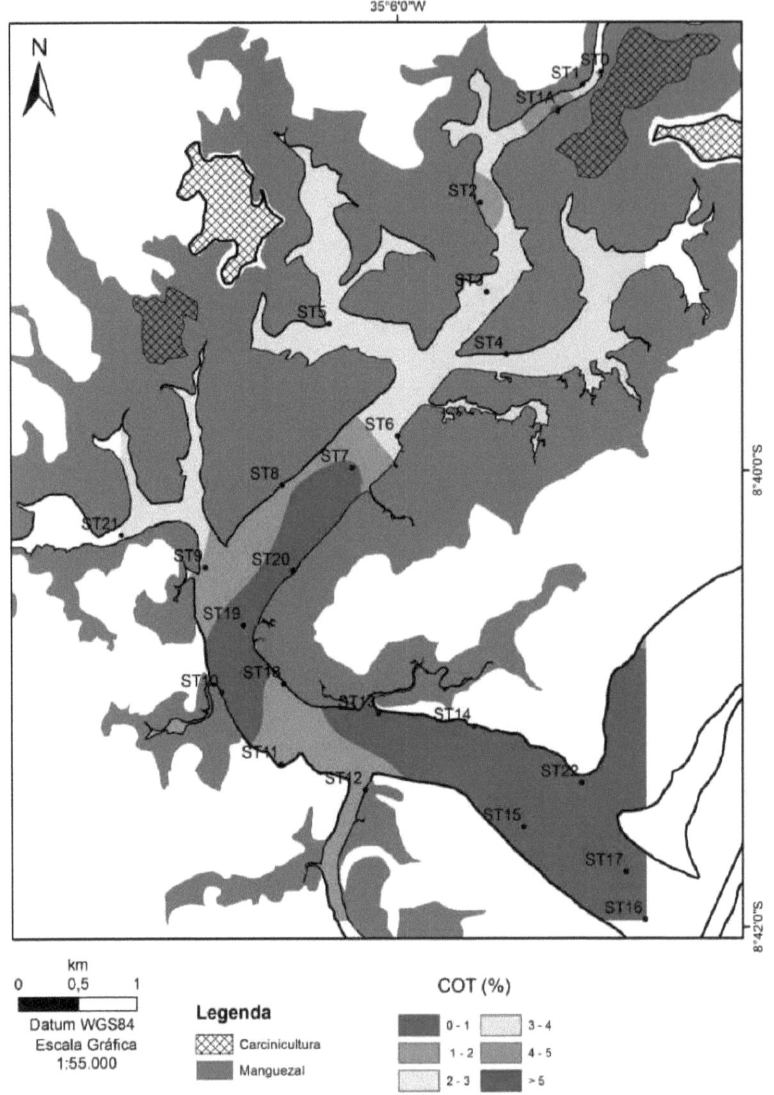

Figura 31: Total Organic Carbon content (%) of sediments in the rainy season (September/2014). Source: Diego Xavier - LABOGEO/UFPE (2016).

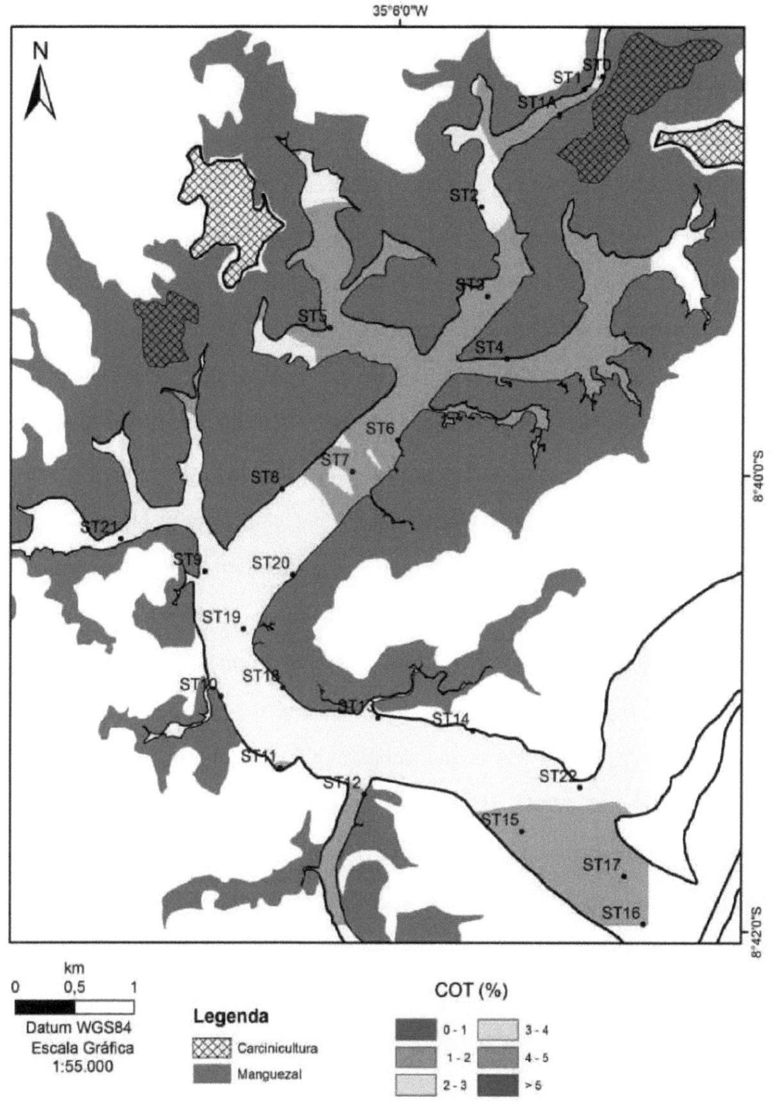

Figura 32: Total Organic Carbon content (%) of sediments in the dry period (March/2015). Source: Diego Xavier - LABOGEO/UFPE (2016).

4.4 Phosphorus (P) content

4.4.1 Total phosphorus content (PT)

Figure 33 shows the map of the total phosphorus (TP) content of sediments from the 2014 rainy season. During this period, the PT contents ranged from 0.02 to 14.30pMol/g, with a predominance of contents between 0.00 and 5.0pMol/g (54% of the samples). The highest PT contents (samples ST12,21,11,06,05) are associated with the highest concentrations of mud (in the samples classified as coarse silt), with the exception of sample ST21, located in the lower estuary and classified as coarse sand.The total phosphorus (PT) contents of the sediments from the dry period/2015 (figure 34) ranged from 0.02 to 23.22pMol/g, with a predominance of contents between 0.00 and 5.0pMol/g (70.84% of the samples). Overall, the PT content decreased compared to the previous period studied, the rainy season/2014, (58.34% of the samples). The higher PT contents (samples ST 01, 04, 05, 20) are associated with the higher concentrations of mud, with the exception of sample ST01, located upstream of the Rio Formoso estuary, which in addition to having considerably higher contents compared to the previous period (7.21 to 23.22pMol/g) was classified as medium sand.

4.4.2 Organic Phosphorus Content

The organic phosphorus (OP) content of the sediments from the rainy season/2014 (figure 35) ranged from 0.01 to 10.17|uMol/g, with a predominance of samples with contents between 0.00 and 5.00 pMol/g (58.3% of the samples). Organic phosphorus (OP) in the sediments from the dry period/2015 (figure 36) showed levels ranging from 0.01 to 17.23|uMol/g, with a predominance of samples with contents between 0.00 (detection limit) and 5.00 pMol/g (83.34% of the samples). They were very similar to the distribution of PT.

4.4.3 Inorganic phosphorus content

The inorganic phosphorus (IP) content of the sediments from the 2014 rainy season (figure 37) ranged from 0.01 to 4.13pMol/g and was very similar to the PT distribution, with samples with contents between

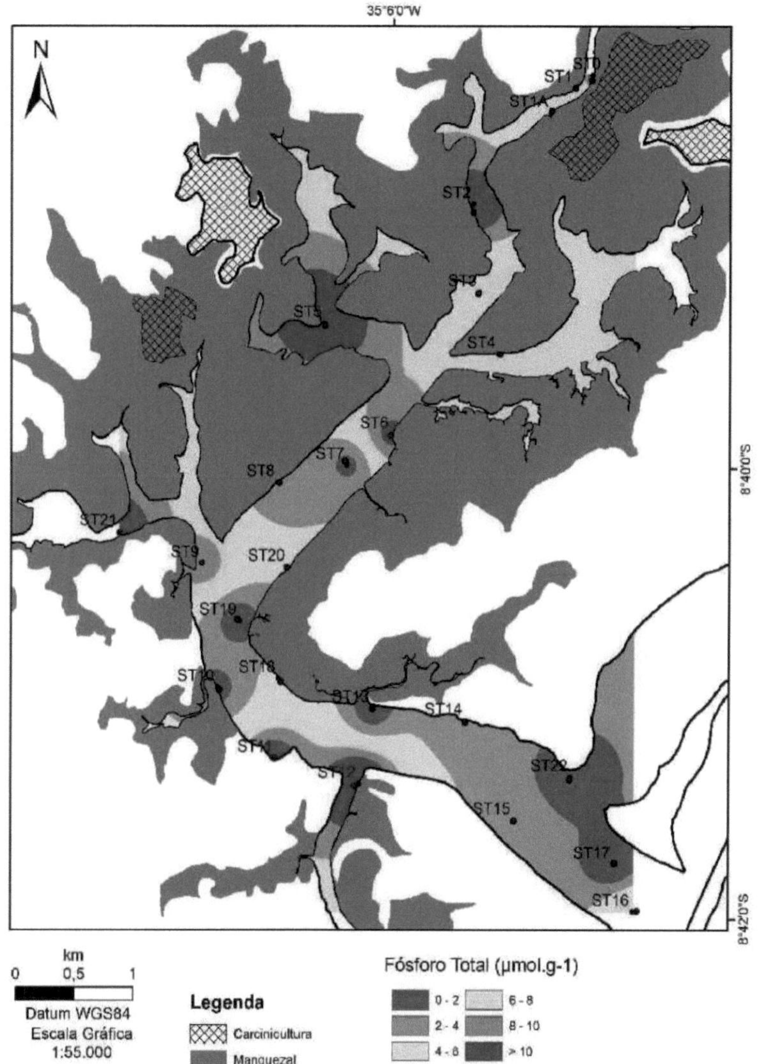

Figura 33: Distribution of the Total Phosphorus content (pMol/g) of the sediments during the rainy season (September/2015). Source: Diego Xavier - LABOGEO/UFPE (2016).

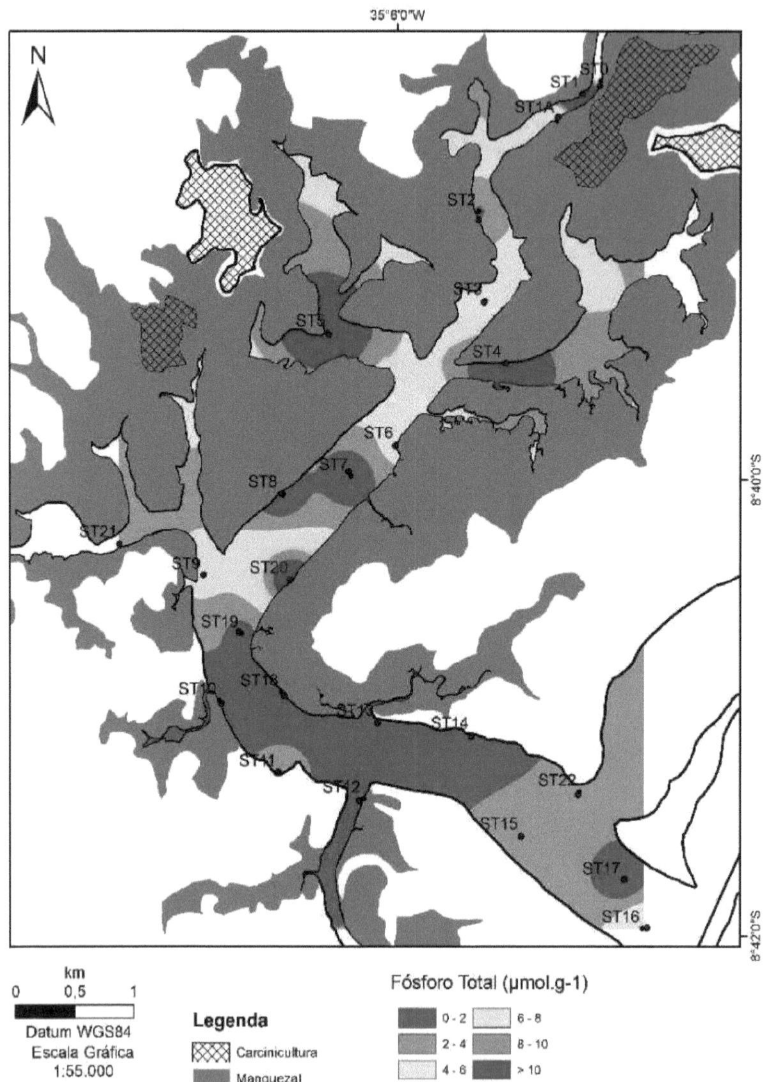

Figura 34: Distribution of the Total Phosphorus content (pMol/g) of the sediments in the dry period (March/2015).

Source: Diego Xavier - LABOGEO/UFPE (2016).

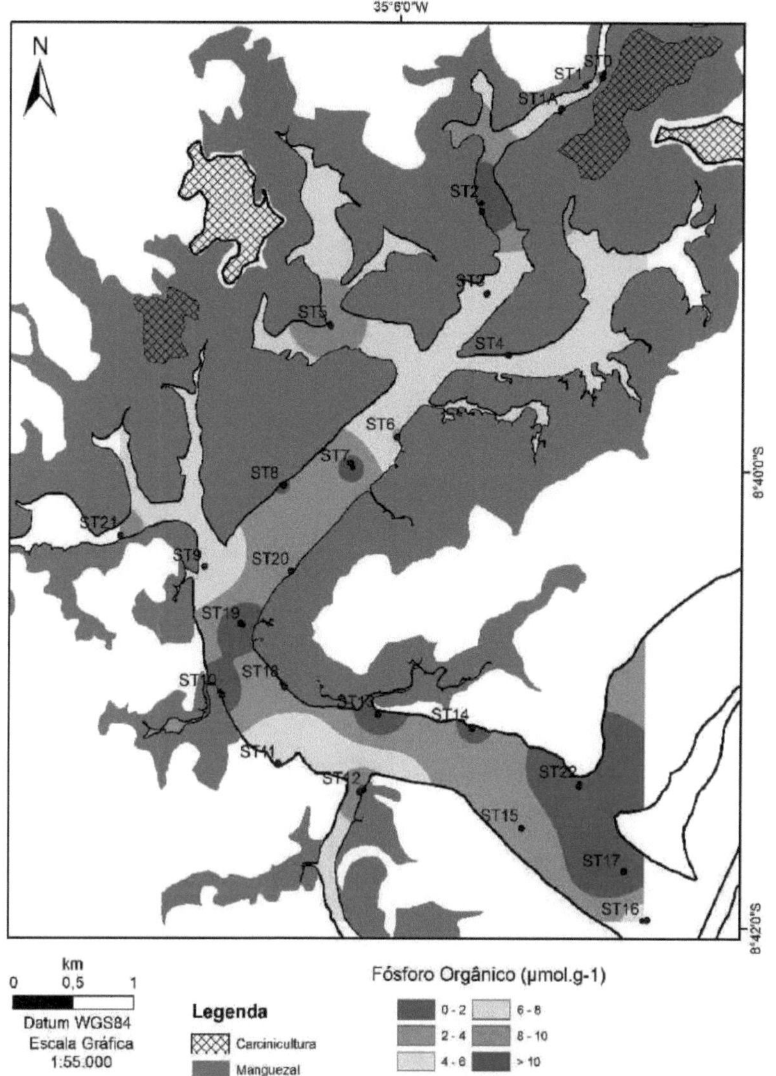

Figura 35: Distribution of Organic Phosphorus levels (pMol/g) in sediments during the rainy season

(September 2014). Source: Diego Xavier - LABOGEO/UFPE (2016).

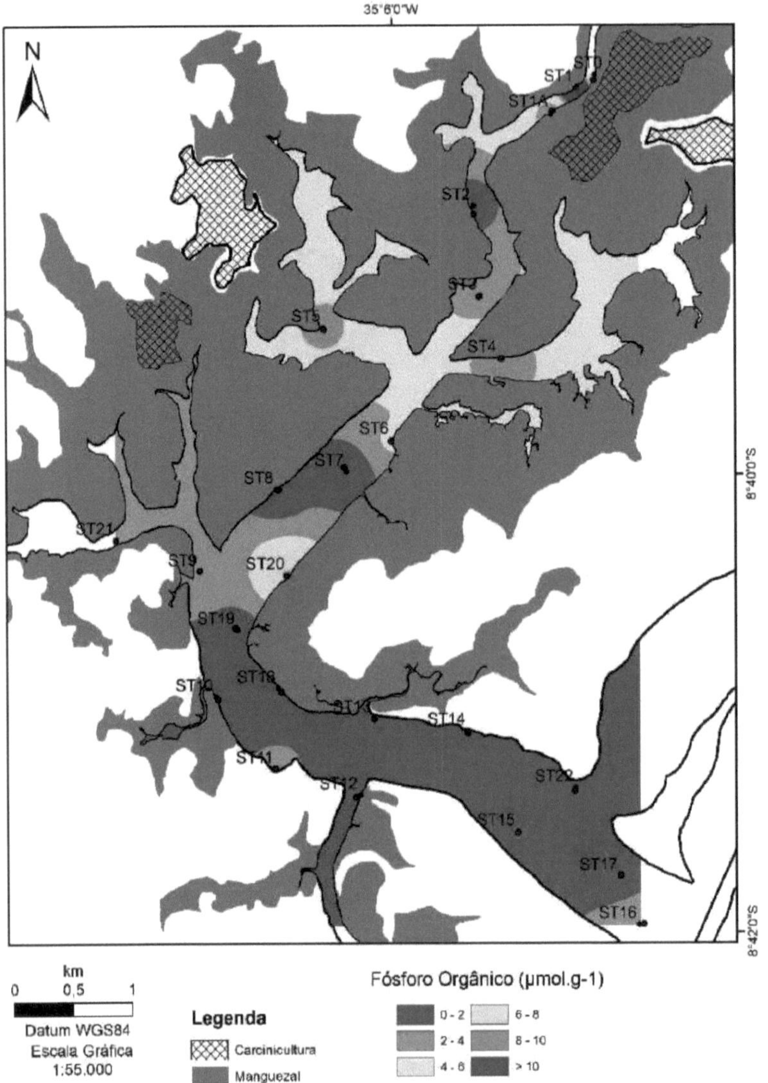

Figura 36: Distribution of the Organic Phosphorus content (pMol/g) of the sediments in the dry period (March/2015). Source: Diego Xavier - LABOGEO/UFPE (2016).

0.00 and 2.00pMol/g (66.7% of the samples). On the other hand, the inorganic phosphorus (IP) content of the sediments from the dry period/2015 (figure 38) ranged from 0.01 to 5.99juMol/g, and was very similar to the PT distribution, with samples with contents between 0.00 and 2.00pMol/g also predominating (70.84% of the samples).

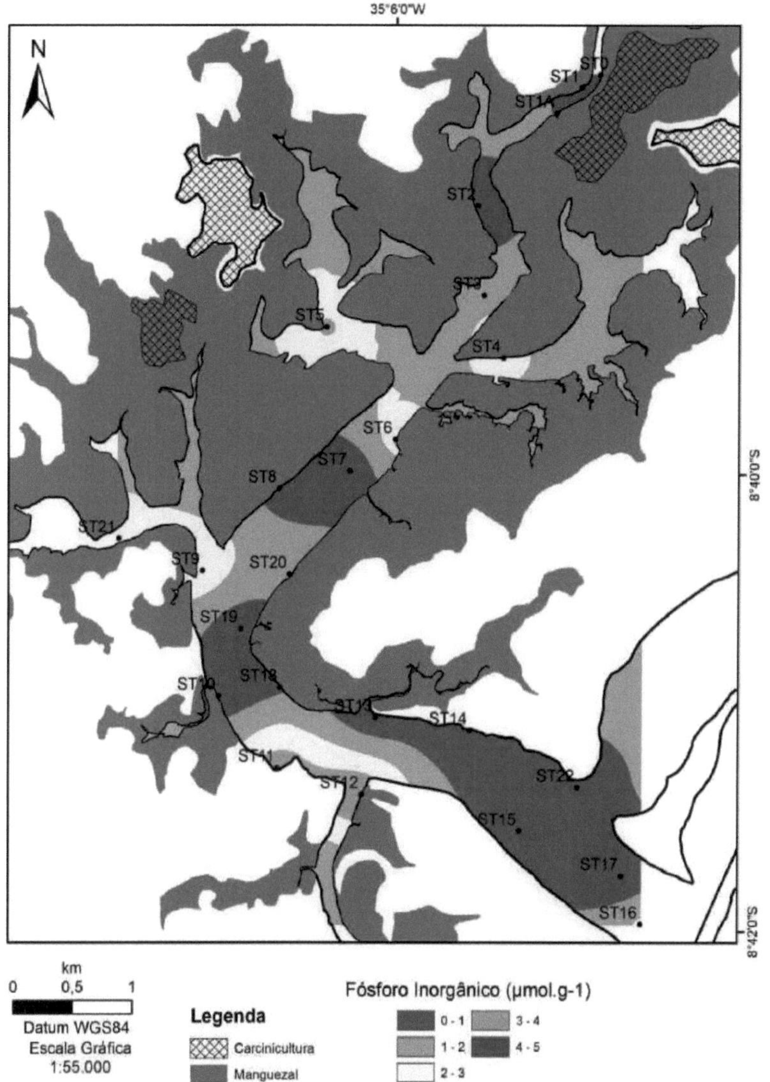

Figura 37: Distribution of inorganic phosphorus levels (pMol/g) in sediments during the rainy season (September 2014). Source: Diego Xavier - LABOGEO/UFPE (2016).

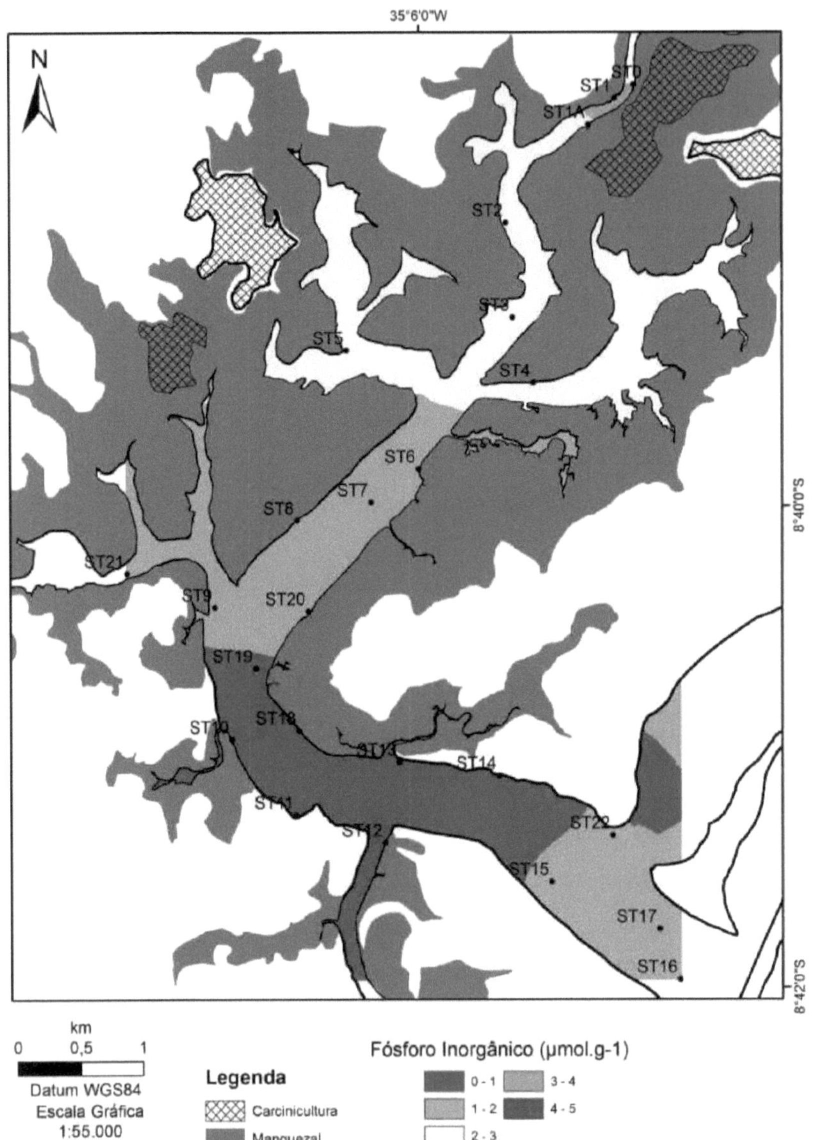

Figura 38: Distribution of inorganic phosphorus levels (pMol/g) in sediments during the dry period
(March/2015). Source: Diego Xavier - LABOGEO/UFPE (2016).

4.5 Elementary ratio C/P

The elemental C/P ratio values of the sediments from the 2014 rainy season (figure 39) ranged from 16.31 to 11337, within the range of values found in the literature for the possible sources of organic matter: phytoplankton (106), leaf and woody tissues of higher plants (300 to 1300 and > than 1300, respectively) and bacteria (7 to 80) (Ruttenberg & Goni, 1997). C/P ratio values indicative of organic matter of predominantly mixed and continental origin were observed in the samples along the estuary. The lower part of the system, on the other hand, showed lower C/P values (< 200), indicative of organic matter of marine and bacterial origin. Overall, there was a dominance of values above 300, indicating a predominantly continental origin for the period studied (rainy/2014).

The elemental C/P ratio values of the sediments from the dry period/2015 (figure 40) varied between 14.14 and 123015. C/P ratio values indicative of organic matter of predominantly mixed and continental origin were also observed in the samples along the estuary. However, lower C/P values (< 200) indicative of organic matter of marine and bacterial origin were observed in some samples from the upper estuary (ST 1, 2, 4 and 5). Overall, there was a dominance of values above 300, indicating a predominantly continental origin, as in the previous period.

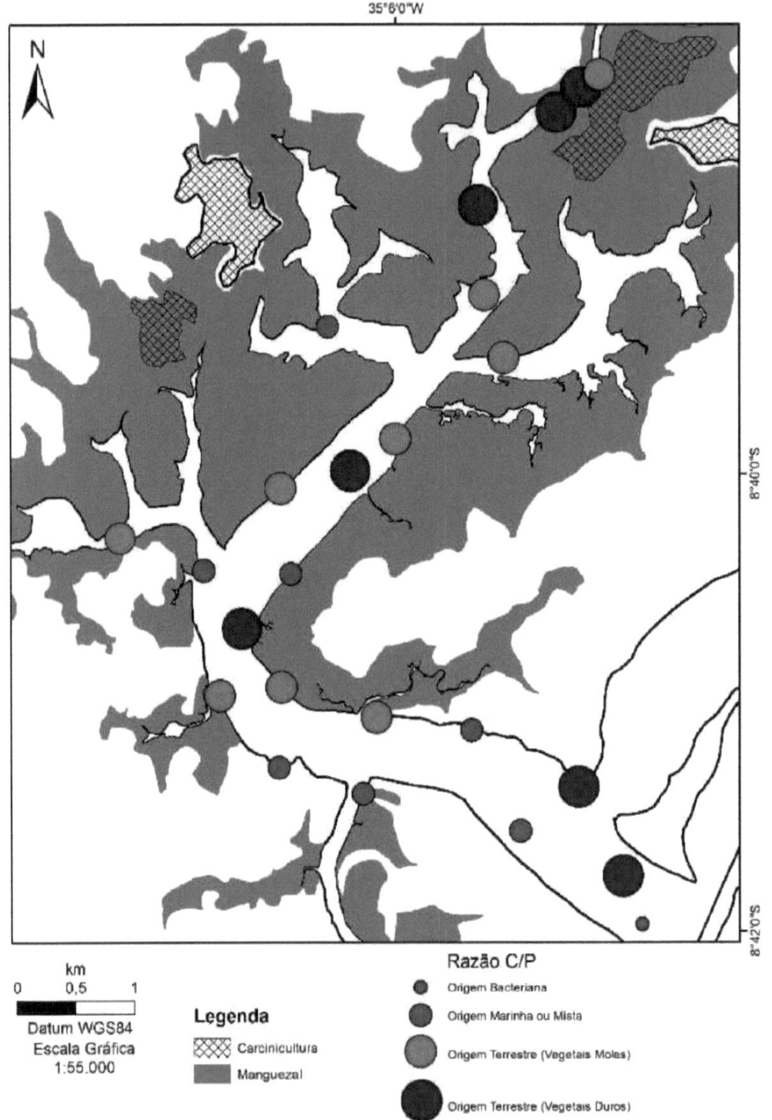

Figura 39: Distribution of the elemental C/P ratio of sediments in the rainy season (September/2014). Source: Diego Xavier - LABOGEO/UFPE (2016).

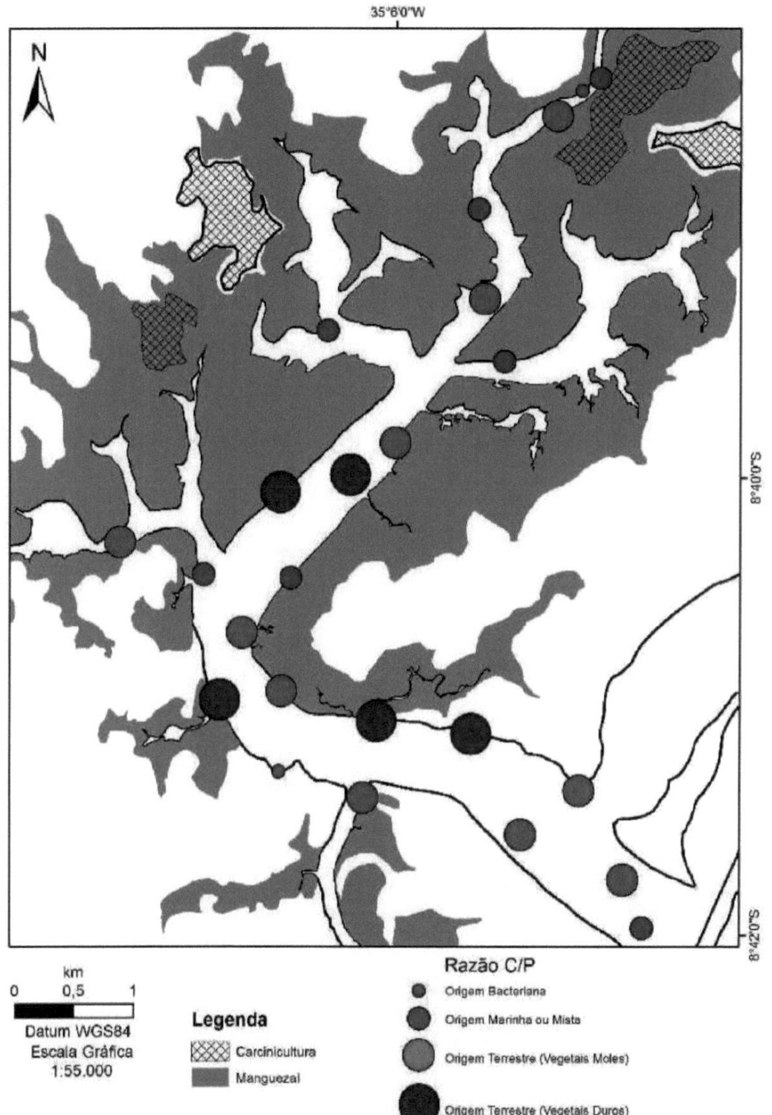

Figura 40: Distribution of the elemental C/P ratio of sediments in the dry period (March/2015). Source: Diego Xavier - LABOGEO/UFPE (2016).

CHAPTER 5

Statistical analysis of parameters: sedimentology and geochemistry

5.1 Cluster Analysis

The results of the cluster analysis for the data collected are shown in Figures 41 and 42 (rainy season/2014 and dry season/2015, respectively). The parameters used in the analysis were: gravel, sand, mud, $CaCO_3$, MOT, TOC, PT, POT, PIT.

The relative sediment facies for the study period allowed three classes to be recognised according to the cut-off level adopted for the rainy season/2014 and two classes and two isolated samples that did not fall into any group at the cut-off level adopted for the dry season/2015. These classes were determined from data obtained through a cross-section of Euclidean distance 4 in the cluster analysis. Their particular characteristics are described in Tables 5 and 6 (rainy season/2014 and dry season/2015, respectively).

Table 5: Characteristics of the sedimentary classes (facies) recognised through the cluster analysis of the sediments in the rainy season/2014.

	Gravel (%)	**Sand (%)**	**Mud (%)**	**$CaCO_3$ (%)**	**MOT (%)**	**TOC (%)**	**EN (%)**	**PO (%)**	**PI (%)**
Easy I	1,14	96,58	2,28	12,45	8,15	1,56	0,02	0,01	0,01
Easy II	0,93	64,34	34,73	10,53	15,73	2,72	0,03	0,01	0,02
Easy III	19,77	67,78	12,45	4,55	16,30	5,33	0,03	0,01	0,02

Table 6: Characteristics of the sedimentary classes recognised through cluster analysis of sediments in the dry period/2015.

	Gravel (%)	**Sand (%)**	**Mud (%)**	**$CaCO_3$ (%)**	**MOT (%)**	**TOC (%)**	**EN (%)**	**PO (%)**	**PI (%)**
Easy I	3,90	95,83	0,27	8,25	3,56	2,07	0,01	0,00	0,01
Easy II	0,11	75,15	24,73	11,6	12,97	3,97	0,03	0,02	0,01
Isolated ST1	10,28	89,54	0,18	3,60	4,03	1,31	0,07	0,05	0,02

Isolated ST IA	1,55	28,05	70,41	17	35,77	6,01	0,02	0,01	0,01

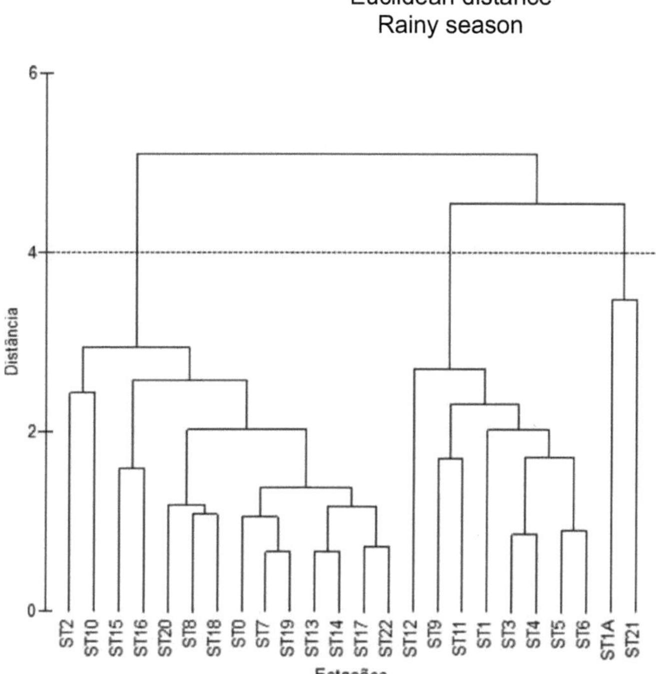

Figure 41: Dendogram - Cluster analysis of sediment in the rainy season/2014.

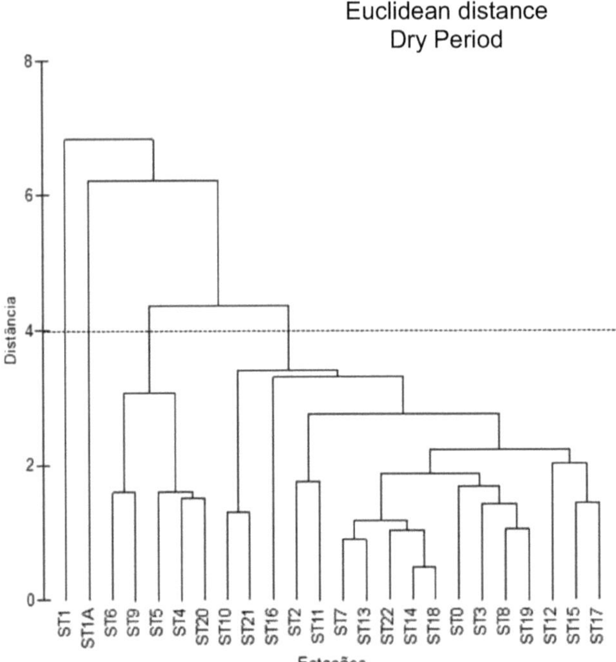

Figure 42: Dendogram - Cluster analysis of sediments in the dry period/2015

The groupings presented were differentiated mainly by the content of sandy sediments, muddy sediments, gravelly sediments, MOT, TOC and inorganic phosphorus (rainy season/2014). In the case of the dry period/2015, the groupings were differentiated by the content of sandy sediments, muddy sediments, MOT and organic and inorganic phosphorus.

Figure 43 shows the estuary subdivided on the basis of these sedimentary facies for the rainy season (September 2014). Facies I is characterised by samples with percentages of the sand fraction greater than 85% with low levels of MOT, organic phosphorus and TOC. The sand fraction averages 96.58%, MOT 8.15% and TOC 1.56%. This facies is predominant in the stations located in the middle and lower sectors of the estuary, with a few stations further upstream, ST 0 and ST 02. Facies II is made up of samples containing the highest levels of mud, MOT, TOC

and organic and inorganic phosphorus. The larval fraction averaged 34.73%, MOT 15.73% and TOC 2.72%, PO 0.01%, PI 0.03%. The facies is distributed in the stations (ST 01, 03, 04,

05, 06, 09, 11 and 12) further up the estuary, with the exception of samples ST 11 and ST12, which are distributed in the lower estuary. Facies III is made up of just two samples (ST 1A and 21) with high gravel fraction contents, 14.08% and 25.46% respectively (average 19.77%), MOT (average 16.30%), low CaCOa contents (average 4.55%) and high TOC content 5.33 compared to the other 2 facies.

Figure 44 shows the estuary subdivided on the basis of these sedimentary facies for the dry period/2015. Facies I is characterised by samples with a higher percentage of sand fraction (average 95.83%) and low MOT content (average 3.56%). Facies II, on the other hand, is characterised by samples with low percentages of the gravel fraction (average 0.11%), high percentages of the sand fraction (average 75.15%), CaCOa and MOT (average 12.97%). This facies is predominant at ST stations 4, 5, 6, 9 and 20 located in the middle sector and at the mouths of the tributaries of the estuary's main channel. In this period (dry/2015), two isolated samples were observed (ST1 and IA). ST1 differs from the others in the following ways: it has a high percentage of gravel (10.28%), it is somewhat similar to facies I in the presence of a high percentage of sand (89.54%), and to facies II in the low content of CaCOa (3.60%), MOT (4.03%) and TOC (1.31%). Sample ST1 A, on the other hand, is characterised by a high percentage of the larvae fraction, CaCOa (17.00%), MOT (35.77%), TOC (6.01%), organic phosphorus (0.05%) and total phosphorus (0.07%).

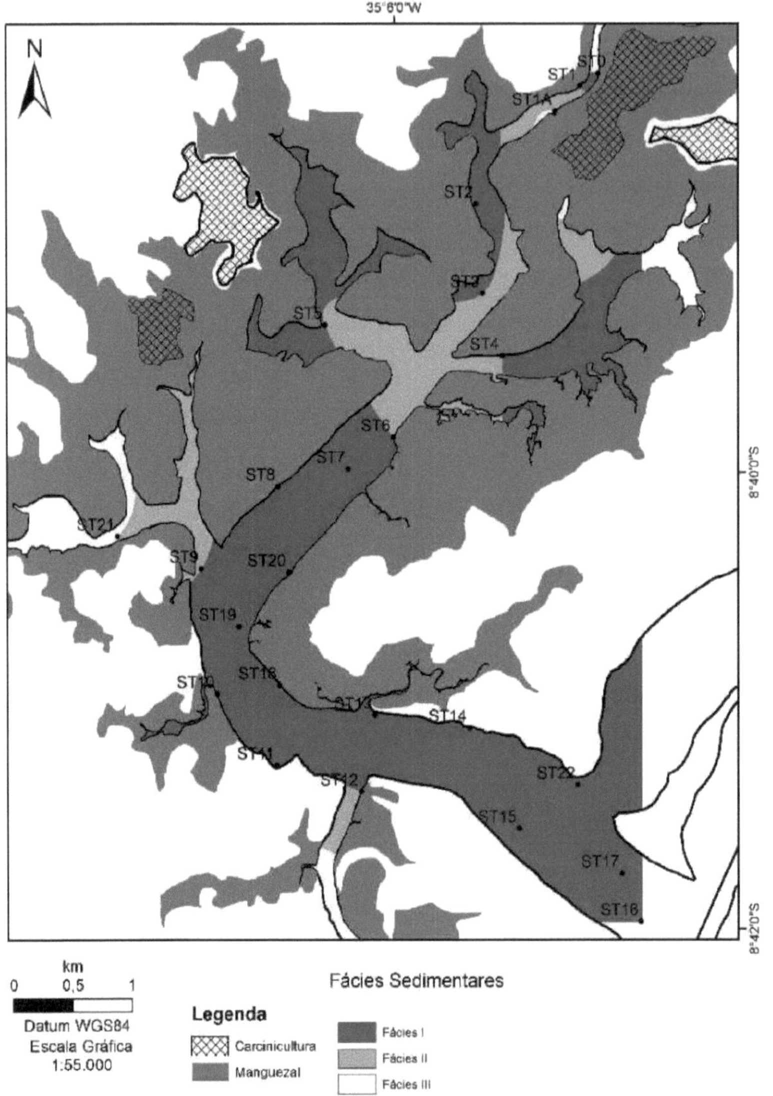

Figure 43: Sedimentary facies defined from the cluster analysis in the rainy season/2014.

Source: Diego Xavier - LABOGEO/UFPE (2016).

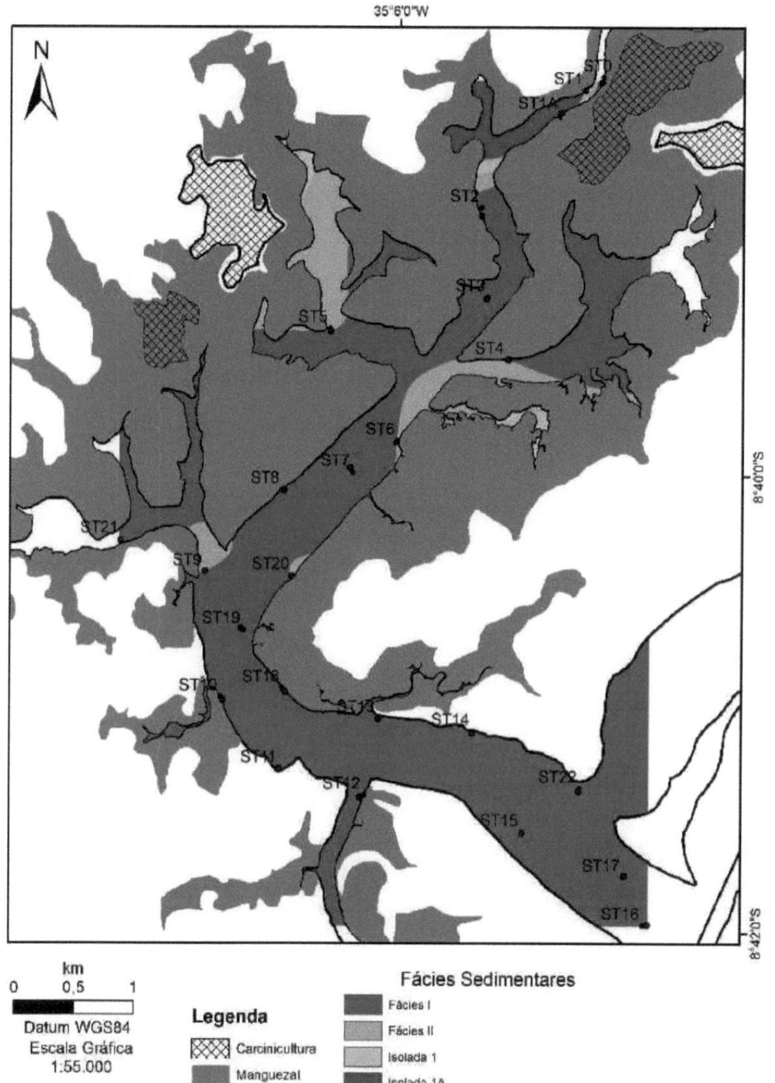

Figura 44: Sedimentary facies defined from cluster analysis in the dry period/2015. Source: Diego Xavier - LABOGEO/UFPE (2016).

5.2 Principal component analysis (PCA)

In the rainy season/2014, the set of main eigenvectors was found to hold 81.6% of the total variation information, with the PCI component showing 63.3% and PC2 18.3% (Table 07). For the dry period/2015, the set of main eigenvectors was found to hold 73.5% of the total variation information, with the PCI component showing 50.5% and PC2 23.0% (Table

08).

The principal components calculated using the sediment PCA are shown in Tables 09 and 10 (rainy season/2014 and dry season/2015, respectively). In the rainy season/2014, the first component (PCI) reflects the sandy sediment content and the second component (PC2) the more gravelly sediment content. Based on this principal component analysis, the samples were grouped into three distinct groups (Figure 45):

- Group 1 brings together the most positive charges in the PCI, corresponding to the samples with sandy sediments.
- Group 2 brings together the most negative charges from PCI and PC2, corresponding to the samples that had fine sediments with the highest levels of organic matter and inorganic phosphorus.
- Group 3 brings together the most negative charges from PCI and the most positive charges from PC2, corresponding to the samples that had fine and gravelly sediments with the highest TOC and MOT contents.

Table 07: Eigenvalues calculated on the basis of the variations used to analyse the sediment in the rainy season/2014.

PC	Eigenvalue	Variation (%)	Cumulative change (%)
1	5,52	63,3	63,3
2	1,59	18,3	81,5

Table 08: Eigenvalues calculated on the basis of the variations used to analyse sediment in the dry period/2015.

PC	Eigenvalue	Variation (%)	Cumulative change (%)
1	4,32	50,5	50,5
2	1,97	23,0	73,5

Table 09: Principal components calculated through the PCA of the sediments in the rainy season/2014.

Variable	PCI	PC2
Gravel (%)	-0,128	0,664
Sand (%)	0,418	-0,083
Mud (%)	-0,349	-0,166
MOT (%)	-0,397	0,061
$CaCO_3$ (%)	-0,127	-0,652
POT (%)	-0,146	-0,081

PIT (%)	-0,452	-0,116
PT (%)	-0,350	-0,117
TOC (%)	-0,408	0,249

Table 10: Principal components calculated using PC A of the sediments in the dry period/2015.

Variable	PCI	PC2
Gravel (%)	0,031	0,222
Sand (%)	0,359	0,265
Mud (%)	-0,360	-0,457
MOT (%)	-0,355	-0,328
$CaCO_3$ (%)	-0,183	-0,250
P Org (%)	-0,523	0,490
P Ino (%)	-0,242	0,174
P Tot (%)	-0,460	0,399
TOC (%)	-0,195	-0,270

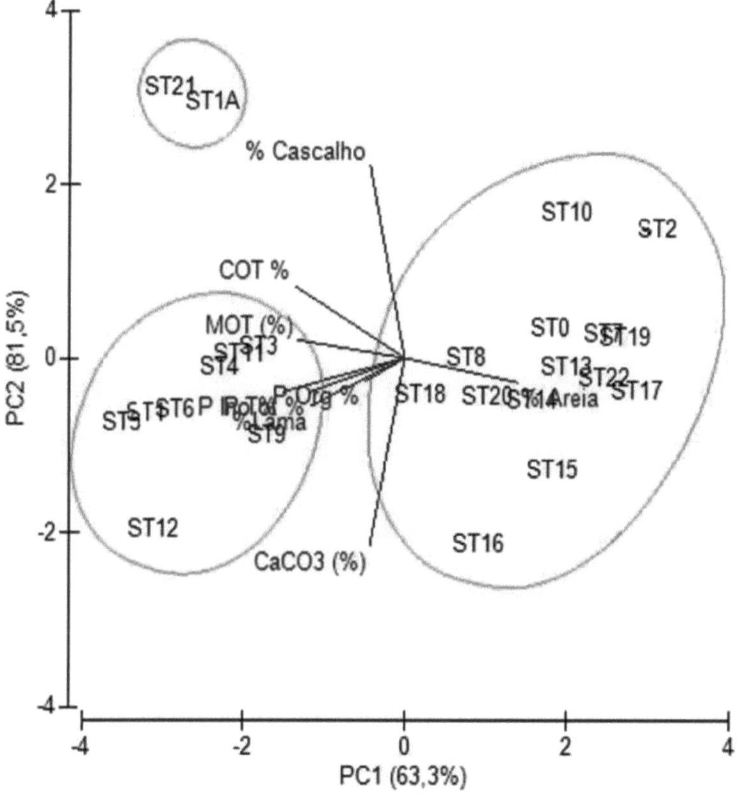

Figura 45: Principal component analysis (PCA) of sediments in the rainy season/2014.
Figura 46:

In the dry period/2015, the first component (PCI) reflects the content of sandy sediments and the second component (PC2) the content of sediments with more organic phosphorus. Based on this principal component analysis, the samples were grouped into two distinct groups and two isolated samples (Figure 46):

- Group 1 brings together the most positive charges in the PCI, corresponding to the samples with sandy sediments.
- Group 2 brings together the most negative charges from PCI and PC2, corresponding to the samples that had fine sediments with the highest levels of organic matter and organic phosphorus.
- Isolated sample ST1 brings together the most negative charges from PCI and the most positive charges from PC2, corresponding to the sample with the highest organic phosphorus and total phosphorus content.
- Isolated sample ST1A brings together the most negative charges from PCI and the most negative from PC2, corresponding to the sample with the highest content of fines and organic phosphorus.

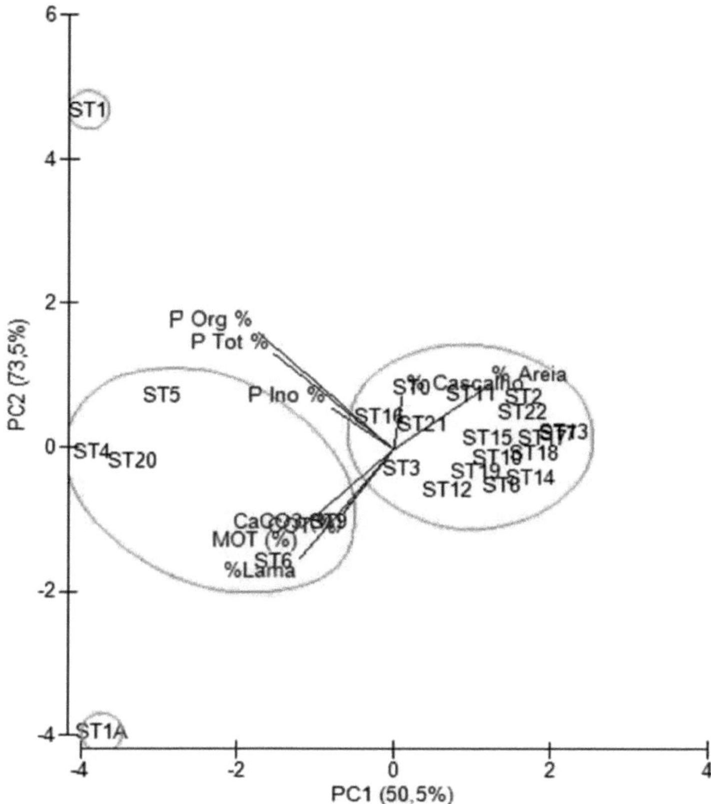

Figure 46: Principal component analysis (PCA) of sediments in the dry period/2015.

CHAPTER 6

Integrated discussion of data

The Rio Formoso estuary has low flows of around 5 m^3/s and a residence time of approximately 11-12 days (LIRA et al., 1979). It belongs to an environmental protection area (APA de Guadalupe), but previous work by Silva (2008) and Santos (2002) indicates that the activities that take place in its surroundings have great potential for environmental degradation. These include: the use of pesticides in sugar cane farming, the dumping of waste and effluents into the Formoso River, the existence of rubbish dumps and shrimp ponds in the upper estuary.

Regardless of the season, the study area has a sedimentary cover made up of terrigenous material (>58.33%), dominated by quartz sands, poorly sorted, with low organic content (predominantly MOT < 10%; TOC < 4% and Porg < 5 pMol/g) and of continental origin (C/Porg >300), associated with a moderate energy environment due to the protection of the reefs at the mouth of the estuary, the medium tidal range and a small net river discharge.

The seasonal sedimentological study found sediments ranging from fine silt to coarse sand. A greater predominance of sandy sediments was observed in the dry period (83.33%) than in the rainy period (70.83%), as well as a greater presence of muddy sediments in the rainy period samples (29.17%) than in the dry period (16.67%). Barcellos et al. (2016) and Oliveira (2014) also found the same seasonal sedimentary trend in the estuaries of the Metropolitan Region of Recife (RMR). Barcellos et al. (2016) in the Jaboatão River estuary found an average increase of 10% in mud for the rainy season. And Oliveira (2014), in the Capibaribe River estuary, found an average increase of 25% in mud also for the rainy season.

For both periods studied, the gravel fraction played little part in the total of the samples, with sediments with gravel fraction contents below 1% predominating (75.00 and 62.50%, wet and dry, respectively). And the highest proportion of the mud fraction with contents above 50% occurred in the upper estuary. With regard to the degree of sorting, poorly to very poorly sorted sediments predominated. The heterogeneity of grain size and, consequently, a decrease in the degree of sorting in the rainy period may be associated with the greater influence of the mixture of river and marine inputs and the increase in mud content for this period. As we move into the dry season, heterogeneity decreases and, consequently, the degree of selection increases for this period. Similar results were found by Silva (2008) in the Rio Formoso estuary, showing a predominance of sandy samples along the estuary, with little participation from the gravel fraction (concentrated at greater depths and close to Pedras beach, in the middle of the estuary), a greater concentration of the mud fraction also upstream of the estuary, and the smallest in the middle and lower stretches associated with the banks of the channel. The author also found a tendency towards poor selection of samples.

However, high levels of gravel were observed in samples ST IA with 14.08% and ST21 with 25.46% (rainy season/2014) and ST1 with 10.28% and ST21 with 16.46% (dry season/2015). It is worth noting that it is not so common to find high levels of this more gravelly material in more internal portions of estuaries, since the hydrodynamic flow in these sectors is lower (Silva, 2008). Therefore, it is suggested that this gravel fraction is not associated with marine biogenic origin, since it did not show a significant correlation with the CaCOa parameter (*rs*= -0.25; p<0.05, rainy season/2014 and *rs*= -0.17; p<0.05, dry season), but rather with plant biogenic origin. In the compositional analysis of the sandy fraction, station ST1A had an average of 50% plant fragments and station ST21 around 30% for the rainy season (September/2014). In the dry season (March/2015), there was a decrease in the content of plant fragments, with station ST21 showing around 20% of these fragments. On the other hand, it was not possible to verify the plant fragment content for station ST1A in the dry period because it did not have enough material in the selected fractions (0.500 mm and 0.250 mm) to carry out the proposed analysis.

In the geochemical study, the calcium carbonate content, regardless of the season, indicates the occurrence of lithoclastic sediments, i.e. with contents below 30% according to Larssoneur et al. (1982). It is possible to observe a gradient of CaCOa increasing towards the sea (figures 24 and 25). In the zone between the reefs and the mouth of the estuary itself, the percentage of carbonate is higher, reaching values of up to 57.40% for the rainy season and 45.30% for the dry season in the sample from ST 16 associated with medium sand according to Shepard's textural classification (1954) for both seasons. This was the sample with the highest marine biogenic sediment content (average of 60%) and the highest B.M. index for both the wet and dry periods. Lira et al. (1979) also observed this gradational behaviour for CaCOa in the Rio Formoso estuary. However, they observed values of up to 14%, much lower than those found in this study. Lira et al. (1979) also observed that there is no source of carbonate for the estuary from the surrounding land since the area is represented by granitic crystalline rocks, some remnants of the Mesozoic volcanism that occurs in the Cabo region and sediments from the Barreiras Group. Similar results with a predominance of lithoclastic sediments were also found for other estuaries on the Pernambuco coast by Alves (2016), Barcellos et al. (2016), Oliveira et al. (2014), Alves et al. (2013) and Oliveira et al. (2013). Alves (2016) in the Goiana River estuary found values between 0.90% and 27.10%, Barcellos (2016) in the Jaboatão River estuary an average of 17.40%. Oliveira (2014) in the Capibaribe River estuary found a variation between 3.70% and 79.10% for the rainy season and 2.04% and 76.10% for the dry season. Alves et al. (2013) in the Jaboatão River estuary obtained an average value of 17.40% in November 2010 and 15.50% in May 2011. Oliveira et al. (2013) in the Capibaribe River estuary observed a range from 6.30% to 19.00% in July 2010 and from 2.50% to 48.70% in September 2011.

With regard to seasonality, 66.67% of the samples from the dry period/2015 showed a reduction in their carbonate percentage compared to the previous period studied. There was also a considerable increase for sample ST1A, located upstream of the estuary (1.9% to 17.0%). CaCOa levels are usually more associated with sediments with higher levels of biogenic material from marine sources. However, ST1A (dry period/2015) showed a considerable increase in its percentage, which was not expected. As it is located in the upper estuary and has no source of carbonate from the surrounding land, this could be an indication of a possible external contribution to the area. Since no fragments of marine biogenic origin were identified in the compositional analysis of the sandy fraction. Alves (2016) also observed samples with a high CaCO.3 content (19.14% to 24.88%) at more internal stations of the Goiana River estuary (PE) adjacent to the shrimp farming areas.

Regarding the total organic matter content, regardless of the season, there was a predominance of values below 10% with averages of 8.15% and 6.88% (rainy and dry, respectively) associated, in general, with sandier sediments located in the central region towards the lower estuary (mouth). Seasonal variation was observed, with higher levels of organic matter in the rainy season. As for spatial variation, there was an increase from the lower to the upper estuary (figures 26 and 27). They showed positive correlations with the percentage of mud *(rs*= 0.81; $p<0.05$ and *rs*= 0.87; $p<0.05$, rainy and dry, respectively), with TOC *(rs*= 0.85; $p<0.05$ for the rainy period/2014 and *rs*= 0.68; $p<0.05$ for the dry period/2015) and with the phosphorus content (total, organic and inorganic) in both periods (appendices a and b). However, values above 25% were found in the sample from ST1 (29.57% in the rainy period) and ST1A (35.77% in the dry period) located in the upper estuary adjacent to the shrimp pond area. Relatively high levels were also found in samples from ST4, ST5, ST6 and ST21 (17.87%, 19.77%, 18.47% and 17.67%, respectively for the rainy season and 16.17%, 13.73%, 15.57% and 4.67%, respectively for the dry season). According to the classification of sediments proposed by Naumann (1930, apud Esteves, 1998), sediments with MO concentrations of less than 10% are of the mineral type, while sediments with more than 10% MO are said to be organic. Oliveira & Matos (2007) associate the high organic matter content with possible external sources and/or anthropogenic influences. Therefore, the explanation for these high values for a low-flow estuary such as the Rio Formoso is due to the fact that they are under anthropogenic influence (shrimp ponds and year-round sewage) in addition to the longer residence time and mangrove areas present in the estuary. The samples from ST1 and ST1A, which are under great influence from carciniculture due to the fact that they are located adjacent to the tanks. Samples ST4, ST5 and ST6, on the other hand, are influenced by the contribution of the tributaries Rio Porto da Pedra, Rio dos Passos and Rio Lemenho which, especially during the rainy season, receive nutrients from the surrounding areas and enrich the estuary. As well as sample ST21, which *in* the rainy season receives a greater influence from open-air sewage from the municipality of Rio Formoso and from urban rubbish dumped on the estuary shore (Lira, 2010). Silva et al. (2009) found values close to those obtained in the present study for organic matter, observing values between 2.7% and 17% for the Rio Formoso estuary. The highest value measured in this study is associated with the sample located near ST5. These authors, however, did not analyse samples near ST1 and ST1 A. High MOT levels and greater seasonal variation in the rainy season were found by Oliveira (2014), who observed MOT contents of up to 30.30% in the Capibaribe River estuarine system during the rainy season, which he associated with natural sources of material from the Pina Mangrove and terrigenous fluvial inputs from the Capibaribe River. NORONHA et al. (2011) found MOT contents of up to 25.60 per cent in the surface sediments of the Timbó River in Paulista (PE) during the rainy season.

In general, the percentage of TOC, regardless of the season, also shows low to medium levels (< 3.00%) and is associated with sandy sediments, distributed throughout the estuary from upstream to the mouth. For the rainy season, values between 0.11 and 7.00% were found, with an average of 1.55%, and for the dry season values between 0.11 and 6.01%, with an average of 2.65%. The highest TOC content was associated with sample ST1A for both the rainy (September/2014) and dry (March/2015) periods (7.00% and 6.01%, respectively), located in the upper estuary near the shrimp ponds. Seasonal variation was observed, with higher levels of total organic carbon in the dry season samples.

In terms of spatial variation, there was an increase from the lower to the upper estuary (figures 28 and 29). According to Siqueira et al. (2003), the distribution of organic carbon is influenced by local effluents, water circulation in the system and physical-chemical parameters. Therefore, a possible explanation for this behaviour of increased TOC in the transition to the dry season may be linked to the system's water renewal capacity, which may be lower in summer than in winter and, consequently, preserving and accumulating more TOC in the dry season.

In addition to the positive correlation with organic matter, TOC showed a positive correlation with the percentage of mud for both periods ($rs = 0.74$; $p < 0.05$ for the wet period/2014 and $rs = 0.58$; $p < 0.05$ for the dry period/2015). Oliveira et al. (2014) also observed similar values for organic carbon content (0.16 to 8.99% for the dry period and 0.06 to 10.99% for the rainy period), with the low content also associated with sandy sediments and the high content with muddy sediments with a high organic matter content. A similar result for organic carbon content in estuaries was found by Barcellos et al (2016) in the Jaboatão River estuary. The author observed values ranging from 0.08 to 7.07% (average of 2.48%) and 0.03 to 8.51% (average of 2.79%) in November 2010 and May 2011, respectively. However, Alves (2016) observed lower TOC values (0.00 to 2.34%) for the Goiana River estuary. As did Barcellos (2005) in the estuarine - lagoon system of Cananéia- Iguape (SP) (values < 2.00% in most of his samples).

Like TOC, phosphorus in the three fractions (PT, POT and PIT) is spatially associated with the highest sludge concentrations (in samples classified as coarse silt) (rs = 0.72; 0.69;

0.73; $p < 0.05$, respectively), with the exception of sample ST21, located in the lower estuary classified as coarse sand according to Shepard (1954) in the rainy period/2014 and sample ST1, located in the upper estuary of the Rio Formoso near the shrimp ponds. This sample, in addition to having a considerable total phosphorus content in the rainy season, had an increase of 222.05% in the dry season (7.21 to 23.22|uMol/g) and was classified according to Shepard (1954) as medium sand. Total phosphorus content varied from 0.02 to 14.30pMol/g with an average of 5.36pMol/g in the rainy season and from 0.02 to 23.22p.Mol/g with an average of 4.68pMol/g in the dry season. There was a predominance of samples with values below 5.00|j.Mol/g in both periods studied. Where the highest content of total phosphorus and organic phosphorus was in the sample from ST 12 (14.30 and 10.17pMol/g, respectively) located at the mouth of the Ariquindá river and classified as coarse silt according to Shepard's triangular diagram (1954), which may be associated with the input of domestic sewage from the centre of Tamandaré (PE). However, this behaviour was not observed for the dry period, since there was a decrease in the correlation parameters of phosphorus content (PT, POT, PIT) with mud content. It was also observed that the highest total and organic phosphorus content was associated with a sample classified as medium sand according to Shepard's (1954) triangular diagram, ST1, with contents of 23.22 and 17.23p.Mol/g, respectively. Normally, higher contents of total and organic phosphorus are found associated with fine sediments, as observed by Gaspar (2008) and Barcellos (2005). However, in the dry period for this study, higher contents were found to be associated with sandier grains, thus showing that this parameter did not follow a direct relationship with grain size. One possible explanation for this may be related to the location of the sample, which is adjacent to the shrimp farming area, and the longer residence time of the estuarine waters for this period of the year.

A slight seasonal variation was found, with higher total phosphorus contents in the rainy season samples (figures 30 and 31). This indicates that rainfall had a direct relationship with phosphorus levels, since the flow of fresh water increased phosphorus concentrations for this period. As for spatial variation, an increase was observed in the samples located near the mouths of the tributaries that form the estuary for both periods. Similar results were found by Gaspar (2008). The ST5 sample showed relatively high values for both seasons (12.91 and

13.59pMol/g, wet and dry, respectively). Seasonally, it was also possible to observe a decrease in the correlation values of TOC with the contents of PT, POT and PIT in the transition to the dry period, varying from ($0.65 < rs < 0.77$; $p < 0.05$) to values of $0.29 < rs < 0.32$; $p < 0.05$. One possible explanation for this different behaviour between the seasons could be linked to the change in nutrient source. It is possible that in the rainy season TOC and phosphorus had the same source (freshwater, continental) and in the dry season different sources (input from shrimp farming or greater marine influence).

Barcellos (2005) also observed an association between total phosphorus values and concentrations of finer sediments. Like Berbel (2008), he did not observe a significant seasonal variation in this element, as in the present study.

The elemental C/Porg ratio values of the sediments for both periods showed variations within the range of values found in the literature for the possible sources of organic matter: phytoplankton (106), leaf and woody tissues of higher plants (300 to 1300 and > than 1300, respectively) and bacteria (7 to 80) (Ruttenberg & Goni, 1997). C/P ratio values indicative of organic matter of predominantly mixed and continental origin were observed in the samples along the estuary. The southern sector of the system, on the other hand, showed lower C/P values (< 200), indicative of organic matter of marine and bacterial origin in the rainy season. And in the dry season, in the samples from the upper estuary (ST 1,2, 4 and 5). A possible explanation for the low C/P ratio values found in the rainy season in the lower estuary may be related to the greater marine influence observed in analysing the sandy fraction and the greater hydrodynamics (Silva, 2008) to which the region is subject. Thus, as organic carbon is associated with finer sediments, in larger particles such as sands, this carbon and phosphorus are probably removed by the bottom currents in contact with the sediment. Only then would a differential enrichment of phosphorus in relation to carbon be reflected in the data obtained, possibly associated with anthropogenic inputs. As the C/P ratio is directly related to TOC concentration and inversely related to phosphorus concentration, the lowest C/P ratio value for the period was observed in the ST 16 sample. Since the sample from ST16 has a high phosphorus value for the region (7.06 pMol/g), this is possibly related to some phosphate input from agriculture. Gãchter & Mayer (1993) point out that, in eutrophic lake environments, it is common to observe C/P ratio values below 106. According to Ruttenberg

& Goni (1997) and Ruiz-Femández et al. (2002) the causes for these low values are: a reflection of the dominance of bacterial biomass; sedimentary organic matter being rich in refractory organic phosphorus compounds in the overlying water column and sediments; the preferential regeneration of C and relative enrichment of phosphorus, due to the transformation of suspended C into gaseous forms of decomposition and; the process of phosphorus adsorption by clays, or even iron oxides, which could protect it from mineralisation. For the low C/P ratio values found in the dry season in the upper estuary, it is believed that they may be related to phosphate inputs from shrimp farming due to waste feed and shrimp excretions in the ponds (FIGUEIREDO et al., 2005), plus the longer residence time of the water in the system at this time of year. This would result in a longer time for water renewal and, consequently, greater retention of organic material in the sediments of the upper estuary. Barcellos et al. (2005) found similar C/P ratio values indicative of a predominantly mixed and continental origin for samples from the Cananéia-Iguape estuarine-lagoon system (SP).

With regard to the sub-environments formed according to the parameters grouped in the cluster analysis, the estuary was subdivided into 3 facies for the rainy season. And 2 facies and 2 isolated samples for the dry period. With seasonality, there is an increase in facies I (70.83%) and a decrease in facies II (20.83%), reflecting the estuary with sandier sediments, with less organic matter and behaving as a retainer of nutrients in the dry period in the upper estuary samples. In this period, facies III disappears and two isolated samples appear, ST1 and ST1A. ST1 differs from the others in the following ways: it has a high percentage of gravel (10.28%), it is somewhat similar to facies I in that it has a high percentage of sand (89.54%), and it is similar to facies II in that it has a low content of CaCOs (3.60%), MOT (4.03%) and TOC (1.31%). Sample ST1A, on the other hand, is characterised by a high percentage of the mud fraction, CaCOa (17.00%), MOT (35.77%), TOC (6.01%), organic phosphorus (0.05%) and total phosphorus (0.07%).

An anomalous behaviour is clear for the sample located adjacent to the shrimp ponds in the upper estuary, ST1A. This was classified as sandy clay according to Shepard (1954) in the rainy season, similar to the sample from ST21, which also had its sedimentary and geochemical characteristics altered in this period due to a greater continental influence (the sample is located closest to the municipality of Rio Formoso) associated with river inputs due to rainfall. In the dry period, it behaved in isolation, taking the sample with the highest percentage of the muddy fraction (70.41 %), the highest percentage of MOT (35.77%), the highest percentage of TOC (6.01%) and a considerable percentage of CaCO.3 (17.00%). In this case, it shows that the estuary in this region and for this period remained a retainer of nutrients, related to the longer residence time of the waters in summer.

CHAPTER 7

Final Summary

When all the aspects analysed are taken into account, the conclusion is that the sediments that make up the Rio Formoso estuary are predominantly sandy and poorly sorted for both the winter and summer seasons. This is due to the moderate energy in the system due to the protection of the reefs at the mouth of the estuary, the medium tidal range and the small net river discharge. The samples that show a better degree of selection are associated with sandy sediment deposits and in the vicinity of the mouth of the Rio Formoso estuary (an environment with greater marine influence and, consequently, greater hydrodynamic energy). The predominance of angular grains in the morphoscopic analyses generally indicates a low-energy environment. And an important temporal variation in terms of grain size was observed, with higher mud content being found in winter as in other estuarine systems in PE (Jaboatão and Capibaribe).

Regardless of the season, there was a predominance of lithoclastic sediments (terrigenous influence and low levels of organic matter, with percentages of calcium carbonate of less than 30% and MOT of less than 10%. In general, the values found for organic carbon were below the warning value (10%), which represents the possibility of causing damage to the environment. However, some stations showed high values of organic matter (areas where pollutants can accumulate adjacent to the shrimp ponds, in the upper estuary), higher than the contents of estuaries in the metropolitan region of Recife, which are highly anthropised.

A seasonal variation in geochemical parameters was observed for the area studied, with higher calcium carbonate and organic matter contents in the rainy season. In the dry season, there were higher levels of total organic carbon. In general, the values found for the phosphorus species did not vary significantly from season to season. The elemental C/P ratio values of the sediments in both seasons indicated a predominantly mixed and continental origin of the organic matter in the samples throughout the estuary. There was little variation in behaviour from one period to the next. In winter, the low values indicated a marine and bacterial origin in the lower estuary and in summer in the upper estuary.

The distribution of sediment and organic matter indicates that the system seems to act as a retainer for material of continental origin, especially in the upper estuary. This is probably associated with the longer residence time of the water in the system in summer. At the same time, the lower C/P ratio values (< 200) indicate a marine character in the upper estuary during the summer, allowing us to affirm that the Rio Formoso estuary also functions as an importer of sedimentary material from the inner shelf.

Therefore, given all the assumptions presented and analysed, it was found that the Formoso River estuary, in the state of Pernambuco, is subject to the anthropogenic actions that surround it. It has shown anomalies in the spatial distribution of organic matter in the areas adjacent to the shrimp ponds and under the influence of sewage from the region's urban areas.

REFERENCES

AGUIAR, V.M.C., 2005. **Spatial and temporal variation of phosphorus and lead biochemical characteristics and transport of properties in the Santos/São Vicente Estuarine System and in the southern portion of the Cananéia - Iguape Lagoon Estuarine Complex (São Paulo).** PhD Thesis. Oceanographic Institute. University of São Paulo. São Paulo. 243 p.

ALVES, C. S., 2016. **Current sedimentation and geochemical study in the Goiana River estuarine system (PE-PB).** Master's thesis. Federal University of Pernambuco. Recife. 78 p.

ALVES, T.M.F; BARCELLOS, R.L; FLORES MONTES, M.J.2013. **Distribution of biodetrital carbonate and total organic matter (TOM) in the estuarine sediments of the Jaboatão River (Pernambuco, Brazil).** In: Congress of the brazilian association of quaternary studies (abequa), 14. Natal. Abstract. CD - ROM.

AMARAL, R. F., 1992. **Analysis for the use and conservation of the Rio Formoso coastal plain - with emphasis on geology and geomorphology.** In: Pires & Filho Advogados Associados. Costa Dourada Project, Recife.

BARCELLOS, R.L 2005. **Distribution of sedimentary organic matter and current**

sedimentary process in the Cananéia Iguape Estuarine Lagoon System, SP. PhD Thesis Oceanographic Institute of USP 187p. 2vs.

BARCELLOS, R. L. et al. 2005. **Distribution and characteristics of sedimentary phosphorus in the Cananéia-Iguape estuarine-lagoon system, São Paulo State, Brazil.** *Geochimica Brasiliensis,* 19(1): 22-36.

BARCELLOS, R.L. & FURTADO, V.V. 2006. **Organic Matter Contents and Modern Sedimentation at São Sebastião Channel, São Paulo State, South- Eastern Brazil.** *Journal of Coastal Research, N.* SI: 39: 1073-1077.

BARCELLOS, R.L.; CAMARGO, P.B.; GALVÃO, A. & WEBER, R.R. 2009. **Sedimentary organic matter in cores of Cananéia-Iguape lagoonal-estuarine system, São Paulo State, Brazil.** *Journal of Coastal Research,* Special Edition n° 56: 1335-1339.

BARCELLOS, R. L.; MONTES, M. J. F.; ALVES, T. M. F.; CAMARGO, P. B. 2016. **Modern sedimentary processes and seasonal variations of organic matter in an urban tropical estuary, Jaboatão River (PE), Brazil.** Journal of Coastal ResearchjcR, v. 75, p. 38-42.

CARREIRA, R.S. & WAGENER, A.L.R. 1998. **Speciation of sewage derived phosphorus in Coastal sediments from Rio de Janeiro,** Brazil. *Marine Pollution Bulletin,* vol. 36, n° 10, p. 818-827.

COMPESA, 2006. **Preliminary Environmental Report Water supply and sewage systems for Tamandaré and the tourist areas of Rio Formoso and Praia dos Carneiros** - update. 227p.

CONDEPE, 1992. **Rio Formoso.** Municipal Monographs, Recife, v.2, 1992. 173p.

CONDEPE/FIDEM, 2006 **Rio Una, GL 4 and GL 5.** Recife: Pernambuco State Planning and Research Agency. 85 p.

CPRH - State Agency for the Environment and Water Resources, 2001. **Socio-environmental diagnosis of the Guadalupe environmental protection area (APA - Guadalupe)** - Recife.

CPRH - State Agency for the Environment and Water Resources, 1998. **Diagnosis Socio-environmental** overview **of the Guadalupe environmental protection area (APA - Guadalupe).** Recife.

CONTI, L.A. & FURTADO, V.V. 2006. **Geomorphology of the São Paulo State continental shelf.** *Revista Brasileira deGeociencias,* 36(2):305-312.

DATTA, D.K.; GUPTA, L.P. & SUBRAMANIAN, V. 1999. **Distribution of C, N and P in the sediments of the Ganges-Brahmaputra-Meghna river system in the Bengal basin.** *Organic Geochemistry.* 30: 75-82.00

ESTEVES, F. A., 1998. **Fundamentos de Limnologia.** 2ª Ed. - Rio de Janeiro: Interciência. Pag 53.

FAIRBRIDGE, R. W., 1980. **The Estuary: its Definition and Geodynamic Cycle.** In: Chemistry and Biogeochemistry of Estuaries, E. Olausson & I. Cato (Eds.) pl-35, Interscience Publication, John Wiley and Sons, New York.

FIGUEIREDO, M. C. B. et al. Environmental impacts of the discharge of effluents from shrimp farming into inland waters. **Engenharia Sanitária e Ambiental,** v.10, n°2, p.167-174. 2005.

FILIPELLI, G.M., 1997. **Controls on Phosphorus concentration and accumulation in oceanic sediments.** Marine Geology, 139: 231-240.

FOLK, R. L. & W. C. WARD. 1957. **Brazos River Bar:** Study of the Significance of

Grain Size Parameters. *Journal of Sedimentary Petrology,* 27 : 3-27.

FREITAS, R.C.; BARCELLOS, R.L.; PISETTA, M.; RODRIGUES, M. & FURTADO, V.V. 2008. **The Valo Grande Channel and siltation in the Cananéia-Iguape estuarine-lagoon system, São Paulo State, Brazil.** *In:* BRAGA, E.D.S (Ed.). Oceanography and Global Change. III° Brazilian Oc.
Symposium. Iª Ed., São Paulo: Ed. IOUSP. p 732-744.

FURTADO, V.V. 1995. **Quaternary sedimentation in the São Sebastião Channel.** *Publ. Esp. Oceanographic Institute* n°l.

GASPAR, L.F. 2009. **Analysis of phosphorus concentration in sediments of the Botafogo and Carrapicho rivers, in the estuarine system of the Santa Cruz channel, Itamaracá, PE.** Master's thesis. Department of Oceanography. UFPE. 77p.

GAUDETTE, H.; MULLER,G.; STORFFERS, P. 1974. **An inexpensive titration method for the determination of organic carbon in recent sediments.** Journal of Sedimentary Petrology, v. 44, n. 1, p. 249-253, 1974.

HILTON, M.J. 1995. **Sediment facies of an embayed costal sandy body,** Paikiri, New Zealand. Journal of Coastal Research, 11: 529-547.

HUANXIN, W.; PRESLEY, B.J & ARMSTRONG, D. 1994. Distribution of sedimentary phosphorus in Gulf of Mexico Estuaries. *Marine Environ. Research'.* (37):375-392.

KRUMBEIN, W.C. SLOSS, L.L. 1963. **Stratigraphy and sedimentation.** San Fransisco : W.H. Freeman and Company.

LAMB, A L.; GRAHAM, P.W. & LENG, MJ. 2006. **A review of Coastal palaeoclimate and relative sea-level reconstructions using §^{13}C and C/N ratios in organic matter.**

Earth-Science Reviews: 75: 29-57.

LIMA, A.F. 1998. **Quaternary sedimentation in Enseada de Fortaleza, north coast of the state of São Paulo.** Master's thesis. São Paulo IOUSP. 138p.

LIMA FILHO, M.F. 1998. **Structural and Stratigraphic Analysis of the Pernambuco Basin.** São Paulo, 1998, 139 p. Thesis (Doctorate in Geosciences) - Institute of Geosciences, University of São Paulo.

LIRA, L.; M.C. ZAPATA, & V.G. FONSECA. 1979. Aspects of the dynamics of the Rio Formoso estuary, PE. **Cadernos Omega da Universidade Federal de Pernambuco 3** (1/2): 133-156.

MALUQUES, M. M. de. 1995. **Current Sedimentary Dynamics in the Coves of the Ubatuba Region,** São Paulo State. *Bolm Inst. oceanogr., NA?>,* n°2: 111-122.

MALUQUES, M. M. de. Considerations on the Bottom Surface Sediments of Ilha Grande Bay, State of Rio de Janeiro. Master's dissertation. Oceanographic Institute - USP, São Paulo, 139 p. 1987. MAHIQUES,

MAHIQUES, M. M.; M. G. TESSLER; A. HOSHIKA; Y. MISHIMA; K. SUGUIO & K. KAWANA. 1997. **Infra-annual Variations in the Characteristics of the Organic Matter from Bertioga Channel, Southeastern Brazil.** 6* Congress of the Brazilian Association on Quatemary Research. Abstracts, Curitiba, ABEQUA, pp. 94-98.

MAHIQUES. M. de; M. G. TESSLER; V. V. FURTADO. **Characterisation of Energy Gradient in Enclosed Bays of Ubatuba Region, South-Eastern Brazil.** Estuarine, Coastal and Shelf Science, 47, 431-446. Academic Press. 1998.

MANSO, V. A. V., et al., 2006. **Erosion and progradation of the Brazilian coast: Pernambuco.** Marine Geology and Geophysics Laboratory - LGGM. Recife: University

Press.

MARTINELLI, L.A., et al., 2009. **Unravelling environmental issues with stable isotopes.** São Paulo, Oficina de Textos, 2009. 143p.

MARTINS et al., 2016. Depositional evolution in a lagoonal estuarine system under a port influence in Northeast Brazil. Journal of Coastal Research, V. 75, p. 83-87.

MEYERS, P. A. 1997. **Organic Geochemical Proxies of Palaeoceanography, Palaeolminologic and Palaeoclimatic Processes.** *Organic Geochemistry* 27, 213-250.

NIMER,E. 1977. **Climate.** In: Geography of Brazil: Northeast Region. Rio de Janeiro: IBGE, v. 2, 1977. pp. 47-84.

NORONHA, T. J. M. et al. 2011. Evaluation of heavy metal concentrations in sediments of the Timbó River Estuary, Pemambuco-Brazil. **Archives of Marine Sciences.** Fortaleza, v. 44, n. 2, p. 70-82, 2011.

OGRINC, N.; FONTOLAN, G.; FAGANELI, J. AND COVELLI, S. 2005. **Carbon and nitrogen isotope compositions of organic matter in Coastal marine sediments** (Gulf of Trieste, N Adriatic Sea): indicators of sources and preservation. *Mar. Chem.:* 95: 163-181.

OGRINC, N. & FAGANELI, J. 2006. **Phosphorus regeneration and burial in near-shore marine sediments** (the Gulf of Trieste, northem adriatic Sea. Estuarine, Coastal and Shelf Science: 67: 579-588.

OLIVEIRA, T. S. et al. 2014. Current sedimentary process and distribution of organic matter in the estuarine system of the rivers Capibaribe, Beberibe and Pina basin, Recife - PE. **Revista de Integrated Coastal Management,** 14 (3):399-411.

OLIVEIRA, T. S. 2014. **Current sedimentary process and distribution of organic matter in the estuarine system of the Capibaribe and Beberibe rivers and the Pina**

basin, Recife - PE. Master's thesis. Postgraduate programme in Oceanography. Federal University of Pernambuco - PE, 112 p.

OLIVEIRA, G. D.; MATTOS, K.M.C. 2007. Environmental impacts caused by the shrimp industry in the municipality of Nísia Floresta **(RN).Revista Ibero Americana de Estratégia,** 6 (2): 183-188.

PAIVA, et al., 2008. **Structure and organisation of shallow-water ichthyofauna in a tropical estuary.** Brazilian Journal of Zoology. V. 25, n. 4, p 647 -661.

PARDO, P. et al., 2004. **Shortened screening method for phosphorus fractionation in sediments a complementary approach to the standards, measurements and testing harmonised protocol.** Analytica Chimica Acta, v. 508, p. 201-206.

PETTTIJOHN, F.J., 1975. **Sedimentary Rocks.** 628p., Harper & Row, 3. ed., New York, NY, U.S.A. ISBN: 0060451912.

PRITCHARD, D. W., 1967. **What is an Estuary:** Physical Viewpoint. In: Estuaries. G. H. Lauff (Ed.) American Association for the Advancement of Science, n 83, Washington D. C.

RASHID, M. A. 1985. **Geochemistry ofmarine humic compounds.** New York, Springer-Verlag. 300p.

REGAZZI, A.J. 2000. **Multivariate analysis, lecture** notes INF 766, Department of Computer Science, Federal University of Viçosa, v. 2.

RILEY, J. P.; CHESTER, R., **1978. Chemical oceanography.v.7.** Academic press: new york

RUIZ-FERNÁNDEZ, A.C.; HILAIRE-MARCEL,C.; GHALEB, B. & SOTO- JIMENEZ, M. 2002. **Recent sedimentary history of anthropogenic impacts on the Culiacan River**

Estuary, NW Mexico: geochemical evidence from organic matter and nutrients. *Environmentalpollutiorr.* 118: 365-377.

RUTTENBERG, K. C. & M. A GONI. 1997. **Phosphorus Distribution, C:N:P Ratios, and ô^{13}C in Arctic, Temperate and Tropical Coastal Sediments:** Tolls for Characterising Bulk Sedimentary Organic Matter. *Marine Geology,* vol. 139, -1/4. P. 123-145.

SANTOS, F.M.; LESSA, G.C.; LENTINI, C.A.D. & GENZ, F. 2011. **Comparative study of the geomorphological characteristics and sedimentary fill of six large Brazilian estuaries.** XIII° ABEQUA, UFRJ, Búzios (RJ).

SANTOS, M. M. F. 2002. **Environmental impacts on the Formoso River estuary from the confluence of the Arinquidá/Formoso rivers, Tamandaré (PE).** Master's dissertation. UFPE. Recife.

SETUR/CPRH, 2011. **Guadalupe environmental protection area:** booklet 3, analysing the conservation unit. Recife. 206p.

SHEPARD, F.P. & D. G. MOORE. 1954. **Sedimentary Environments Differentiated by Coarse Fraction Analysis.** *Buli. An. Assoe. Petrol. Geol.,* 38(8):1792-1802.

SILVA, J. B., et al., 2009. **Geomorphological classification of estuaries in the state of Pernambuco (Brazil) based on satellite images.** In: XII Congresso da Associação Brasileira de Estudos do Quaternário - ABEQUA; IV Congreso Argentino de Cuatemário Y Geomorfología; II Reunión sobre el Cuatemário de América del Sur, La Plata, Argentina; IV Congreso Argentino de Cuatemário Y Geomorfología; II Reunión sobre el Cuatemário de América del Sur, La Plata, Argentina, 21 to 23 September 2009.

SILVA, J. P. 2008. **Sedimentological, hydrodynamic, bathymetric and water quality studies aimed at the evolution and associated environmental aspects of the Rio Formoso estuary - PE.** PhD thesis. Postgraduate in Geosciences, Centre for Technology and Geosciences, Federal University of Pernambuco, 146p.

SIQUEIRA, G. W. 2003. **Study of the levels of heavy metals and other elements in surface sediments from the Santos estuarine system (Baixada Santista - São Paulo) and the Amazon continental shelf (Northern Continental Margin).** PhD Thesis, IOUSP, 327p.

SOUZA, M.F.L., et al., 2012. **Biogeochemical and physical carbon cycling and interactions between compartments in Todos os Santos Bay.** Revista virtual de química, 4(5), 556-582, ISSN: 1984-6835

STEIN, R. 1991. **Accumulation of Organic Carbon in Marine Sediments.** Results from the Deep Sea Drilling Project/ Ocean Drilling Programme. *In:* Bhattacharji, S; G. M. Friedman; H. J. Neugebauer; A. Seilacher (Eds.), ***Lecture Notes in Earth Sciences,*** Vol. Springer, Berlin, 217pp.

TUNDISI, J.G.; TUNDISI, T.M., 2008. **Limnology.** São Paulo: Oficina de Textos.

TYSON, R.V., 1995. **Sedimentary Organic Matter: Organic facies and palynofacies.** 615p., Chapman & Hall, London, U.K. ISBN: 978-94-010-4318-2.

WILLIANS, et al.,1976. **Forms of phosphorus in the surficial sediments of Lake Eire.** Journal of Fishers Research Board Canadian, v. 33, p. 413 - 429.

XU, S.; GAO, X.; LIU, M. & CHEN, Z. 2001. China's Yangtze estuary. II. **Phosphorus and polycyclic aromatic hydrocarbons in tidal flat sediments.** *Geomorphology:* 41: 207-217.

Printed by Books on Demand GmbH, Norderstedt / Germany